Other books about the home by Stanley Schuler

How to Design and Build a Fireplace
How to Fix Almost Everything
The Wall Book
The Floor and Ceiling Book
Homeowner's Guide to Home Remodeling
The Complete Book of Closets and Storage
The Homeowner's Minimum Maintenance Manual
All Your Home Building and Remodeling Questions Answered
5000 Questions Answered about Maintaining, Repairing and Improving Your Home
The Complete Terrace Book
How to Build Fences, Gates and Walls
Encyclopedia of Home Building and Decorating
The Householder's Encyclopedia
The Homeowner's Directory
Illustrated Encyclopedia of Carpentry and Woodworking Tools, Terms and Materials

Table of Contents

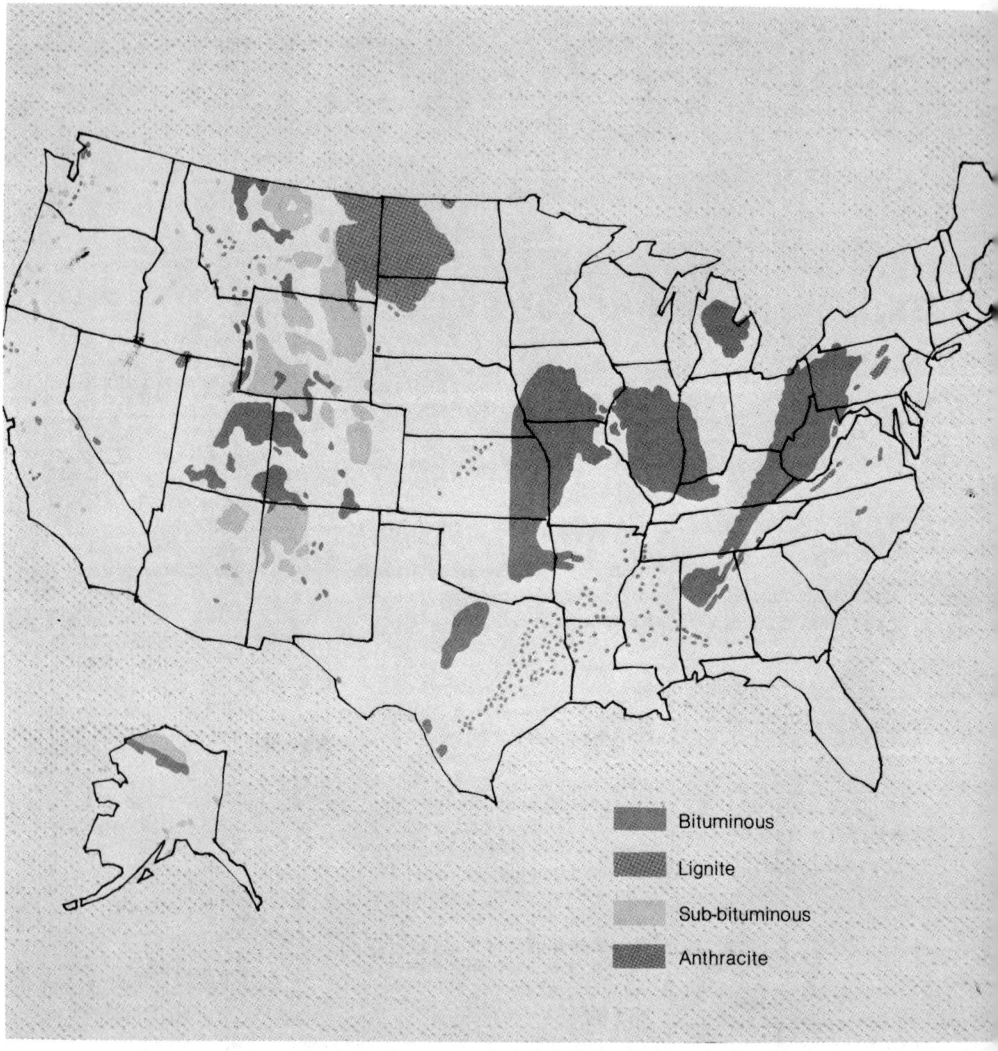

Coal is mined almost everywhere in the United States, and there is lots and lots of it. (National Coal Association)

Chapter 1
Should you switch to coal?

Except for the home owner and small business owner who burn natural gas, we're all looking for alternative ways to heat our homes and small businesses. At the moment, natural gas is comparatively cheap. But sooner or later, the pendulum will swing and gas will lose its price advantage. Then everyone will be seeking alternative heat sources.

Will coal be the fuel we turn to?

Until the end of the 20th century, ninety percent of the energy we used for all purposes–heating homes, making steel, producing electricity, running factories and railroads–was furnished by coal. Then oil began to inch its way into the picture and by 1920 coal furnished only seventy percent of our total energy. By the mid Thirties, the figure had dropped to forty-five percent, and today it's only about twenty-five percent. This means there is still a substantial amount of coal being used, but only about two percent of the total mined is sold at retail and winds up in the furnaces of homes and small businesses. So if there's any expectation that you and your neighbors will be heating with coal in the near or distant future, a dramatic change must take place.

The "experts"–economists, sociologists, planners, etc. in government and out–are agreed this will never happen.

Coal has been billed as the fuel that will save the country from the evil Arabs. The basis for this belief is the fact that the United States has coal coming out of its ears. The U. S. Geological Survey estimates that we have known reserves of about 1.7 trillion tons plus an additional 1.8 trillion tons in unmapped, unexplored areas. That's enough coal to supply all our energy needs for the next century.

But the experts don't believe we're going to return to the day when the great majority of heated small buildings had a coal bin in the basement, because they don't think that home owners and small business proprietors any longer want to be bothered with tending a coal fire. If coal does become the primary source of energy in the United States–and incidentally, the experts no longer think it will be–it will be used to produce electricity or synthetic oil; and our homes will be heated by electricity or synfuel–not by coal. In other words, the experts believe home use of coal will not increase. On the contrary, one of the Department of Energy whiz-bangs in Washington told us it was going to decrease.

If experts hadn't been proved so wrong about so many things, we'd be inclined to accept their judgment. Since World War II, Americans have been so spoiled by completely automatic heating systems that it's hard to believe they will ever willingly or unwillingly return to coal heating–even with automatic stokers that do everything except remove the ash from the furnace.

But the American people are not predictable. And for the past few years they have been appalled, mad, even desperate about the cost of heating their homes. As a result, they are taking action in their own ways. No more waiting for government, industry, the "experts" to come up with an answer. They're looking for and finding their own answers.

The first thing they did, at least in our more heavily forested regions, was to start burning wood in fireplaces and newly purchased stoves. There is no more archaic or laborious way to heat–or partially heat–a house than with wood. But Americans are doing it–without any suggestion or urging from the experts. Here in our little corner of southeast Connecticut we know a dozen or so families that are heating entirely with wood. Even though the cost of firewood has soared–right here in an area that is almost entirely forested, a cord costs ninety-five dollars if you can't or won't cut your own–they're saving money, keeping as warm as toast and laughing at the oil dealers who can't believe what is happening.

Well, if home owners are willing to heat with wood, why shouldn't they be willing to heat with coal?

Now—this is the second step in this transitional period—more and more are. Many of the people who were heating with a wood stove have switched to coal because, for reasons we'll discuss later, they like it better. And the next thing you know, they'll be installing coal furnaces.

The sales of coal-fired residential furnaces and boilers are on the rise. In April of this year one manufacturer told us he had already sold out his entire planned production for 1980. Is this the writing on the wall?

We're not experts. We just put out a magazine telling people how to repair and improve their homes—and being constantly amazed at the things those same people will do when they get mad. We long ago learned that you can't say, "Americans will never do this." Americans do what they damned well please. And if it pleases them to go back to burning coal to keep warm, they're going to do it.

We're not advocating coal heat. Our position is neutral. Our aim in this book is simply to tell you what coal heating is all about. You can make up your own mind when you're finished. You would anyway, no matter how positive or negative about coal we were. That's the American way, the "experts" notwithstanding.

We say we're neutral about coal heat. That's the honest truth. Neither of us is using it at the moment.

Stanley Schuler, aged sixty-five, is thirty-four year old Cary Hull's father. The only direct contact Cary has ever had with coal was about twenty years ago when we occasionally burned cannel coal in our chilly dining room's fireplace. Today, she and her husband Charles own a big old Victorian house with an ancient coal-fired boiler that was at some time in the past converted to oil. The inefficiency of the boiler doesn't bother them much because they used it last year to burn only fifty gallons of oil. Two wood stoves were their main source of heat. But they're tiring of cutting and hauling wood and feeding it into the stoves and they occasionally think about improving their central system. They might convert the boiler back to coal; then again, they might not.

Stanley, of course, grew up with coal heat. His wife Elizabeth did, too. Living in the Philadelphia area, Stanley's family burned hard coal and several times a year a huge coal truck would back up to the house and pour a cascade of rather small, evenly sized shiny black rocks down a steel chute through a basement window into the bin. There was never any question about switching to oil. For one thing, the house was kept warm at modest expense. For another thing, the place was staffed by a German couple, and the man was responsible for keeping the furnace running. Even so, an old Iron Fireman stoker was installed eventually.

In Louisville, Elizabeth's family burned soft coal and her father tended the furnace and took out the ashes himself. He probably disliked the job, especially when he was in his sixties; but he showed nothing but pride in it. When a brother-in-law who had just converted to oil suggested that he follow suit, he snorted at the idea. Coal was the best fuel going, he maintained. But like everyone else in Louisville, he decried the effect soft coal had on the atmosphere. Like all Ohio River towns, it was a filthy city, immersed in smog and sinus trouble. Buildings had to be repainted frequently. And every year the fastidious home owner hired a wallpaper contractor either to clean the wallpaper throughout the house with a dough-like substance or repaper.

Stanley's and Elizabeth's first house in Connecticut was a compact, two-story, six-room Colonial set about one hundred feet back from the road and twenty feet above it. In the basement was an anthracite-fired hot-water boiler. When the coal was delivered, not more than twice a year, the driver and his helper lugged the coal up the hill in large canvas bags and chuted it into the bin through a basement window in back of the house. Even so, the coal cost only about ten dollars (memory says it was less) a ton. At that rate, it cost only fifty dollars to sixty dollars to heat the uninsulated house very comfortably every year—a trifle less than a month's rent.

There was no stoker but it wasn't much work to keep the system running. Before he left for work at 7:30, Stanley shook the fire down and threw in several shovelfuls of coal. He checked it when he got home at 6:00 and maybe added a little more coal. Then he banked the fire when he went to bed. About twice a week he shoveled out the ashes; and when the two big ash cans were full, he carried them up the outside basement stair and the trash collector carried the ashes off to the town dump with the garbage.

Now, forty years and three houses later, would he do it again? Maybe. His present house is five times as big as his first. Last winter, heating it to sixty degrees with a modern three-year-old oil boiler cost fifteen hundred dollars, and he used a wood stove in one end of the house in addition. He's done just about everything he can with insulation, weatherstripping and storm sash to conserve heat. So as the price of oil continues to rise, a point will be reached when something has to be done. His alternatives will be electricity (much too expensive), bottled gas (pretty cheap right now but will it stay cheap?) and coal. But if he went to coal, could he find a dealer with men who would carry it up hill and over a wall to get it into the basement?

Coal has certain advantages as a home-heating fuel. It also has disadvantages. Consider the latter first:

1. The supply is controlled by the United Mine Workers, a once tough but now disorganized (therefore more dangerous) union made up of men who, with considerable reason, distrust and detest everyone, especially the mine owners. True, the UMW represents only part of the coal-mining work force, but when it goes out on strike, it causes vast disruption in the country.

Of course, the union or union members usually give ample warning that a strike may occur; so users of coal have time to hoard before the mines are shut down. Even so, a prolonged strike like that in 1977–78 might mean that many houses are cold.

2. The railroads, which transport most of our coal, are in a sad state. They don't have enough rolling stock to carry the coal if demand suddenly increases. Yet they're reluctant to invest in new stock because there is no certainty that the electric utilities, the biggest by far users of coal, are going to increase their demand for it.

Here again, there is relatively little danger that residential coal customers need ever run short of coal. But we can't say flatly that they won't. Whenever our rail system slows down or comes to a complete halt, everyone in the country is hurt.

3. Although the composition of coal varies greatly, coal by and large is a very serious polluter of the atmosphere. It emits sulfur dioxide, nitrogen oxide, trace elements such as arsenic, lead, mercury and fluorine, and carbon dioxide. These make the air we breathe dirty and hazardous and they kill our crops and forests by acidifying the rain. Furthermore, if the scientists are correct, the carbon dioxide (which is generated by many other things in addition to coal) is creating around the world an invisible high-altitude blanket that is slowly disrupting the world's climate.

Because of these frightening effects, the big utility and industrial users of coal are under orders from the federal and state governments not to burn coal without taking measures to clean the emissions from their smokestacks. At the moment, home owners are not required to take similar steps because, presumably, there are not enough homes with coal heat. In other words, the only thing that might constrain you from burning coal is your own conscience, your desire not to contribute to the pollution problem. But how long will home users of coal go unregulated?

Wood fires pollute the air seriously, too. Vermont's normally crystalline air last winter was thick with smog. Where did it come from? The chimneys of the thousands of homes that are heated by wood.

Imagine, therefore, what our air would be like if millions of home owners started burning coal. It could be very bad indeed and made worse by the fact that the coal smoke and pollutants from residential heating plants are dispersed into the air only twenty to thirty feet above the ground (whereas emissions from utility and industrial smokestacks are a hundred or more feet above the ground).

4. Getting rid of the ashes from millions of coal furnaces would be no joke either. For one thing even without the ashes, town after town throughout the United States is already running out of dump sites. In the second place, when coal ash is soaked with water, it leaches many undesirable elements into the ground and causes a sharp rise in soil alkalinity.

Unlike wood ash, coal ash contributes nothing to the home garden. It has no value for any purpose. The hard cinders from coal used to be used as a base for terraces, walks, sometimes roads. They're every bit as good as the bank-run gravel (which we are running out of) that we use today. But there's a limit to how much more paving we can or should do in the United States.

5. If you're going to burn coal, you have to store it in a dry bin near the furnace (for convenience). That takes up space—roughly forty cubic feet per ton.

If you have a basement, as most houses built before World War II did, coal storage is no problem. But what if your house is built on a slab?

6. In the old days, most houses were located and designed so coal could be chuted directly into the basement bin from the coal truck. If they weren't, no matter: there were plenty of men who were glad to have a job delivering coal in sacks on their backs. Are there such people today? Not many.

So much for the disadvantages of heating with coal. What are the advantages?

1. At the moment at least, it's cheaper to heat with coal than with oil, electricity or wood (that you buy). Natural gas is the only fuel that is substantially cheaper than coal. Propane gas is marginally cheaper. (More about how you figure the cost of heating in the next chapter.)

Admittedly, this situation probably won't hold for long. The prices of all fuels are changing. Because of political pressures, only natural gas has held fairly steady; but it can't do this forever because our supplies are being exhausted. A headline in the Wall Street Journal of June 10, 1980 tells the story succinctly. It said, "Natural Gas Prices Spurt Toward Parity With Oil as Consumers Race to Convert."

In the long run, we have observed, all fuel prices have a way of leveling out; and when this happens, coal, which is a less efficient heat source than gas, oil and electricity, may be a trifle more costly than gas and oil and considerably more costly than electricity. But this is frankly guesswork. No one can predict fuel costs very far into the future.

2. As we said before, our coal reserves are enormous. Assuming the miners and railroads do not cause problems, you and your children are assured of an ample supply for your lifetimes.

3. The more coal we burn, the less dependent we become on the OPEC nations for oil. This, in turn, will greatly improve the United States economy by lowering our rate of inflation, improving our balance of trade, etc.

At the same time, increased use of coal preserves our own dwindling supplies of oil and gas and helps to assure that in time of war we have more of these essential materials.

4. Coal was for so long our main source of energy that everyone connected with its production and consumption was content to sit back and rake in the cash. They devoted little time or money to research; did little to improve mining and transportation methods or methods of burning coal more efficiently. Meanwhile, competing industries were pushing ahead on all fronts.

But now, since the OPEC nations went into action, the situation is changing rapidly. The oil companies and others have moved into coal and together with the government, they are sinking a lot of money into research which should eventually have major effect on coal production and consumption. Coal users are bound to benefit as a result.

True, we do not know in exactly what way they will benefit.

But there's reason to believe that coal-fired heating systems will be vastly improved; and this will inevitably make coal heat even more competitive with gas, oil and electric heat and more attractive to the home owner.

5. Because of the abundance of coal, its use will permit you to return to the days when you could set your thermostat to 68, 70, 72 or 75 degrees without feeling guilty about it. Frankly, to us, this is not a very important argument in coal's favor: we've grown accustomed to living in a sixty degree temperature. But there are countless people, especially the elderly, who really do suffer when the indoor temperature dips below seventy degrees.

We leave it to you to weigh the advantages of coal heat against the disadvantages. But even having done that, you can't come to any real conclusion about whether you should use coal heat until you consider the other alternatives:

Natural Gas. No doubt about it, this is an ideal heat source. But the prices will rise and the supply is dwindling.

Bottled Gas. This is used in several million homes and numerous oil-users in the East are converting to it. But the long-run picture is not bright because it is derived from natural gas.

Oil. Oil has been priced out of the market; and even if the price were to tumble sharply, the government would take steps in the name of national security to limit oil use.

Electricity. If gas is an ideal heat, electricity is the perfect heat. It's one hundred percent efficient; subject to the most exacting control. But unless the utilities can come up with a way to generate electricity on a large scale at low cost, it may always be more expensive than other fuels. No energy-producing industry is having so much trouble with government regulations and environmentalists, and there's no sign that this sorry state of affairs will ease up in the foreseeable future.

Heat Pump. The heat pump is a miraculous machine that produces more BTUs of heat than it uses to make the heat. But it's run by electricity; and despite thirty years of intensive development, it is still expensive to run in the northern half of the country.

Wood. At the moment, United States supplies of wood appear inexhaustible but it's well to remember that in other parts of the world that once had tremendous timber reserves, people are now freezing and dying because they've cut it all. We can trust that won't happen here. Even so, wood has become expensive; many people tire of the work involved in cutting, splitting and handling it; and the environmentalists are getting down on it because of the air pollution it causes.

Solar Heat. Some day solar energy will probably occupy the place that coal did in 1900. But despite the hullaballoo about it, we're a long way from developing solar systems that can substantially lower heating costs in all parts of the country. Furthermore, solar heating as we now know it is best adapted to new houses that are built specifically for it. In order to adapt it to the average existing house, you're going to have to rebuild the house extensively—and at today's building costs, ouch.

Weighed against these heat sources, coal looks pretty interesting, doesn't it?

Chapter 2
Coal facts

Coal, gas and oil are known as fossil fuels because they are dug out of the earth. Coal is the most plentiful–in enormous supply in all parts of the Northern Hemisphere. Formed mainly during the Carboniferous period two hundred and fifty million years ago, it is derived from deep layers of dead plants that have been subjected to the great pressure and heat of the earth's crust. The resulting brown to black, stone-like material is ranked according to the amount of carbon, ash, volatile matter and moisture it contains. The lowest rank–with the least carbon and most volatile matter and moisture–is lignite. The highest rank is anthracite.

The constituents of coal should be understood before you examine the types of coal.

Average Percentage Composition of Coal

Property	Anthracite	Bituminous	Sub-bituminous	Lignite
Moisture	1.4	4.8	18.4	41.5
Volatile Matter	6.5	32.3	33.8	23.0
Fixed Carbon	79.5	51.2	39.0	20.9
Ash	12.6	11.7	8.8	14.6
Hydrogen	2.4	5.0	5.9	6.8
Nitrogen	0.8	1.3	1.0	0.5
Oxygen	3.2	10.3	29.3	46.5
Sulfur	0.8	2.7	0.7	1.7
Sulfate Sulfur	0.02	0.16	0.04	0.24
Pyritic Sulfur	0.35	1.70	0.35	0.68
Organic Sulfur	0.48	0.88	0.32	0.75

Carbon

Carbon is the element in coal—and all other fuels—which gives off the heat. The higher the carbon content, the better the coal's fuel value. This is why anthracite, with a carbon content of almost 80% is prized.

Ash

The ash that is left in a furnace after coal has been burned is not the residue of the carbon (which should burn away completely). It was in the coal as it came from the mine. It is composed of impure, or extraneous, matter that is found in all coal seams.

Some coals contain much more ash than others, and generally these should be avoided because you buy coal by weight and if it has a high percentage of ash, you don't get as much heat value out of it as you do from coal with a lower percentage. However, ash is not all bad because you need some of it in a stove or furnace to insulate the grate and to keep small lumps of coal from falling through the grate before they're burned. Furthermore, the ash content can be reduced by washing coal before shipping it from the mine mouth.

The large, ugly, misshapen clinkers that coal sometimes forms when it burns are attributable to the constituency of the ash in the coal. Clinkers are produced when particles in a coal are heated, melt and fuse together. If the ash has a low-fusion temperature, clinkers occur frequently; if the ash has a high-fusion temperature, they do not.

A so-called free-burning or non-fusing coal is one that doesn't form clinkers; the lumps retain their size and shape until they are reduced to ash. Anthracite is the prime example of a free-burning coal. Bituminous is a clinker-forming coal.

Moisture

Like ash, moisture is an undesirable ingredient in coal. Unfortunately, it's found in all coal in part because the plants from which it's derived grew in bogs and the bogs were later completely submerged in prehistoric oceans. But moisture content varies widely—from as much as 70% in some lignite to as little as 2% in anthracite.

A high moisture content is undesirable because it adds to the weight of the coal you buy, takes the place of the valuable carbon, and cools the fire. Furthermore, during storage, the moisture evaporates and the coal begins to crumble and flake.

Some of the moisture in coal can be—and occasionally is—eliminated by drying the coal before it is shipped to market. This increases the BTU value of the coal per pound. But the drying operation adds to the cost of the coal.

Volatiles

The volatiles, or volatile matter, in coal are the gases which are distilled and driven from it by burning. In a stove or furnace, the volatiles are driven off first; then the carbon starts to burn.

The volatiles consist of combustible hydrogen and carbon and incombustible nitrogen. The sulfur in coal is often considered as an incombustible volatile too, but today is often listed as a separate constituent.

The volatiles are the reason why environmentalists are unenthusiastic about increasing coal consumption in the United States. They contribute to air pollution as well as to the gradual formation in the high atmosphere of a carbon dioxide blanket that is altering earth temperatures. At this time, no one can say for sure which of the volatiles is the most serious—if, in truth, they are serious. But the sulfur in coal is the most worrisome; therefore, the higher a coal's sulfur content, the less popular it is—at least among the major consumers who are banned by law from using high-sulfur coal unless they install extremely expensive and elaborate scrubbers to extract the sulfur before it escapes from the chimney.

Sulfur in coal occurs as particles of pyrite and as organic sulfur.

If the coal is washed—as eastern producers of bituminous coal often do—the pyrites can be reduced in volume by 50% but the organic sulfur is not affected. Washing raises the price of the coal; even so, the coal is probably a better buy than unwashed high-sulfur coal because, quite apart from the fact that it is less of a polluter, it burns with a little less smell and is not so corrosive to the furnace and flue.

Types of coal

Because shipping costs affect coal prices to such an extent, coal dealers don't go out of their way to buy and sell only the best coal. Instead, they buy what they can get easily and cheaply so they can, in turn, sell it cheaply and easily. This is why east coast dealers sell only anthracite coal, mined in eastern Pennsylvania; why mid-western and southern dealers sell only bituminous coal, mined in their backyards (so to speak); why western dealers sell the lignite and sub-bituminous coal mined in their backyards.

In other words, if you're a New Yorker who takes a liking to sub-bituminous coal from Wyoming, you're not very likely going to be able to buy it unless you guarantee to pay for a whole hopper car of Wyoming coal. Similarly, if you live in Denver, you probably can't buy anthracite from Wilkes-Barre. True, this situation may change—in the eastern half of the country at least—if the major users of coal start shipping in whole trainloads of western coal. But don't count on it.

There is, however, a somewhat brighter side to this picture: If you can cut your heating costs substantially by burning a coal that comes from a mine not too far away, you won't be so particular about the kind of coal you buy. We talked, for example, to several western users of sub-bituminous and lignite coal. It didn't bother them in the slightest that these coals contain a lot of moisture and relatively little carbon, because they were readily available and cheap. "So what if we have to buy maybe twice as much coal as we'd need if we burned anthracite," was their attitude. "It still saves us a bucketful of dough."

Coal users in Virginia and Kentucky have the same feeling about using bituminous coal, even though it gives off a lot of sulfur dioxide when burned. (But of course their attitude may change if the government ever decrees that small users of high sulfur coal must either put in scrubbers or switch to low-sulfur coal.)

Be all this as it may, you ought to know what coals are available and how they differ. Here they are:

Peat

Peat is not usually called a coal, though if left in the ground for a few more million years, it would become one. In the U. S. we think of peat only as a crumbly, light brown vegetable material that we mix into the soil and use as a mulch. But in other parts of the world, notably Ireland, peat is dug up and chopped into blocks and used for fuel. It contains less than five thousand BTUs per pound (that is, when burned, it produces less than five thousand British Thermal Units of heat*) and gives off a lot of moisture and smoke.

A handful of the stoves sold in the U. S. are able to burn peat, but no one recommends that you use them in this way because you won't get much heat out of them. Furthermore, you probably can't find any peat that is usable as fuel since Ireland put an embargo on its exportation.

Lignite

Like most coals, lignite occurs in several forms. The lowest form, called brown coal, is found around Dallas and has a BTU content of only five thousand, five hundred per pound. Very little is used in homes. The more common lignite is found in east Texas, the western half of North Dakota and the eastern third of

---

*A British Thermal Unit is the amount of heat required to raise one pound of water one degree F. It is the equivalent of 251.996 IT (International Steam Table) calories or about one-third of a watt hour.

Montana. It is harder and blacker than brown coal and has a BTU content ranging from five thousand, five hundred to eight thousand per pound. Containing 21% carbon, 23% volatile matter and 42% moisture, it isn't really a very good heat source but people living in the regions where it is mined think it's just fine because it sells for about twenty-five dollars a ton. At this price, you can burn an awful lot of it without breaking the bank. But perhaps the best thing about it is that it has a very low sulfur content and a low ash content, so it is not a serious polluter. On the other hand, it has such a low ignition temperature that it may spontaneously burst into flame in storage. It also loses moisture in storage and may therefore crumble and flake. Moral: Buy it in small quantities; use it fast; and watch out that it doesn't flare up in the bin.

Sub-bituminous

In recent years especially, coal in the United States has been grouped under the headings of Eastern coal and Western coal. Eastern coal mined east of the Mississippi River accounts for 46% of our total supply by weight and 70% by heat content. Because of its high heat content and because it is mined right in the heart of our eastern and mid-western industrial complex, Eastern coal is used far more than Western coal. But in the few years since national attention was riveted on environmental matters, Western coal has been gaining a bigger foothold in the market because it generally contains less sulfur than Eastern coal, and also because it's so easy to strip from just below the earth's surface that, despite high transportation costs, it can be delivered to, say, Indiana for a price comparable to that of harder-to-mine Eastern coal.

Sub-bituminous coal, like lignite, is a Western coal (found mainly in Wyoming and northern Colorado). It is harder than lignite, but softer than bituminous coal. It contains between eight thousand and eleven thousand BTUs per pound; 15–30% moisture; but only 0.6–0.7% sulfur, and not too much ash. It burns cleanly and freely (it is subject to spontaneous combustion but less so than lignite). Because of its moisture content, it eventually crumbles in storage.

Bituminous coal

Bituminous is our most widely used coal for all purposes (three hundred and forty thousand homes are heated with it). Most Americans have never seen any other kind of coal. Although lignite and sub-bituminous are softer, bituminous has acquired the name of soft coal to differentiate it from anthracite, which is called hard coal. Huge deposits of bituminous coal are found in many states east and west of the Mississippi, but the most important beds extend from western Pennsylvania and Ohio down through West Virginia, Kentucky and Tennessee into northern Alabama.

Because it's mined in so many places, bituminous coal varies widely. The BTU content of that from the midwest ranges from ten thousand, five hundred to twelve thousand per pound while that from the east ranges from twelve thousand to fourteen thousand per pound. (Most people think anthracite has a higher heat content than bituminous, but in actuality some bituminous is as good as or better than anthracite).

Bituminous is often called a smelly coal because the average bituminous has a higher sulfur content than the other coals. But sulfur content also varies from state to state and even within the same mine seam. You can find bituminous with as little as 0.8% sulfur and as much as 8% sulfur.

Ash and volatile content vary similarly.

By and large, bituminous is considered to be dustier and dirtier than anthracite because it breaks up more in storage and burns with more smoke. But if you see black smoke issuing from a furnace flue, you can blame it more on the furnace operator than on the coal.

One other characteristic of bituminous which differentiates it from anthracite is that, as the coal burns, it tends to form clinkers. Anthracite is reduced mainly to ash.

Coal Heat

Anthracite

When anthracite, or hard coal, was formed, it was subjected to greater pressure than bituminous; consequently it contains a higher percentage of carbon and a lower percentage of moisture than all other coals. It also comes from much deeper in the earth (in eastern Pennsylvania) and is therefore more costly to mine and to buy. But along the Eastern Seaboard it is the coal home and shop owners have always used. There are some one hundred and sixty thousand homes, mostly in the East, burning anthracite today.

Composed of 80% carbon, 7% volatile matter and less than 2% moisture, anthracite is hard to ignite; but once it's going, it burns freely and cleanly and very hot. It doesn't fuse but leaves a substantial ash residue (meaning that it has a high ash content; but no one worries much about this since it generates so much heat). It has a sulfur content of only 0.6%, and theoretically you can rub it without dirtying your hands.

Because of its excellent characteristics, anthracite would be much more widely used if it were not fairly scarce (another reason for its relatively high cost). It is sold almost exclusively through retail dealers for delivery to homes and small businesses.

Cannel coal

Cannel coal is called a freak coal because it exists in small pockets (usually adjacent to veins of bituminous coal) in Virginia, West Virginia, Kentucky, Tennessee, Pennsylvania and Utah. It might also be called a freak because it varies so much that it defies generalization. The BTU content ranges from eight thousand to fourteen thousand per pound. It contains a generally high but very variable percentage of volatile matter. It also has a high ash content, leaving a lot of dust and clinkers.

Cannel coal should never be burned in an enclosed area, such as a stove or furnace, because it burns too high and too hot to be readily controlled. But in a fireplace it's delightful. It burns more actively than any other coal—with about as much flame as a well seasoned chunk of pine. It's only failing is that it frequently explodes—not like a bomb; more like an inch-long fire cracker—and sends small fragments all over the place. Fortunately, these are rarely burning, so there isn't much danger of setting the house on fire; but you do have quite a mess of shards to sweep up the next day.

Cannel coal is sold in bags or boxes of from about forty to eighty pounds in weight. Only a few coal dealers handle it; you're just as likely to find it in hardware stores. In any case, the seller collects a handsome price for it—about two hundred and twenty dollars per ton. Happily it goes a long way because you need only a few chunks to make a beautiful, long-lasting fire.

The stuff comes from the mine in large chunks, which the dealer may break into smaller chunks and which you may then break into still smaller chunks. If you hit it with a dull axe, it splits like wood.

Coke

Coke is made from bituminous coal by baking it at high temperature to drive out all the impurities. After it comes from the oven, it is pure carbon, and when burned, this gives off a vast amount of heat without producing smoke and without leaving ash.

Coke is rarely sold through retail channels because it's used primarily in steel production. We found only one dealer (in Chicago) who sells it—at twenty-five dollars more per ton than he asks for anthracite. This may be just as well because it's questionable whether coke doesn't burn too hot to be used in small furnaces and stoves. Only a few stoves, all made in Europe and not on sale here, are designed for coke.

Coal briquettes

Do not confuse coal briquettes with charcoal briquettes, which are made of wood. Coal briquettes are made by mixing pulverized coal with a clean binder in a mold and squeezing them into very dense, compact chunks. They're fairly widely used in Europe in stoves and until recently were imported into the United States. The only American firm we have found making them is the Hubbard Coal Company (431 Stephenson Hwy., Troy, MI 48084). It turns out lignite briquettes with a BTU of ten thousand per pound and anthracite/bituminous briquettes with a BTU of over fourteen thousand per pound. Both sell through garden centers, grocery stores, etc. for about six dollars for fifty pounds, which puts them at about the same price level as cannel coal. They burn easily, quietly and well with almost no smoke; produce very little ash; and have such a low sulfur content that it might as well be called non-existent.

Hubbard also produces compressed coal logs made of the same coals for use in fireplaces. A six and one half pound log that burns for about eighteen hours costs two dollars.

(Another kind of coal log is sold by St. Bernard Coal Co., 2601 Upper River Road, Louisville, KY 40206. It is actually an attractive cardboard tube filled with seven pounds of two by three-inch lumps of high-BTU, low-sulfur bituminous coal.)

What does coal cost?

As the foregoing indicates, coal is a variable product; and it's pricing is even more variable. This makes it extremely difficult—even impossible—to answer the question you most certainly are asking: What does it cost?

The more desirable coal is (high BTUs, low ash or low sulfur), the more it costs. The more difficult it is to mine, the more it costs. The more it is treated—graded into sizes; washed to remove ash and sulfur; oiled to make it cleaner to handle—the more it costs. The smaller the quantity you buy, the more it costs (as a rule). The more the dealer has to do to deliver it (if he has to carry it in sacks or a wheelbarrow, for instance), the more it costs.

If you live near a mine, coal is so cheap that you can hardly afford not to heat your home with it. At any distance from the mine, the story is obviously quite different. This spring in New England, anthracite was selling for about one hundred and five dollars a ton; in Chicago, bituminous was selling for eighty-six dollars; in Des Moines, sub-bituminous was selling for sixty dollars.

But even at these prices, coal currently is a good bargain if you are presently heating with oil, electricity or wood.

Comparing costs

To make a fair comparison of fuel costs, you have to take into consideration the efficiency of the heating equipment in which the fuel is burned. These efficiencies vary somewhat between one make of equipment and another, but are generally figured as follows:

Electric resistance heaters	100%
Gas-fired furnace	75%
Oil-fired furnace	65%
Coal furnace	60%
Wood furnace	50%

The BTU values of the different fuels are:

Electricity	3413 BTUs per kilowatt-hour
Natural gas	1000 BTUs per cubic foot
Oil (No. 2 fuel oil)	138,000 BTUs per gallon
Coal	12,500 BTUs per pound
Wood (seasoned hardwood)	7950 BTUs per pound

To determine how the heat value of a ton of coal compares with the heat value of the other fuels, you must first multiply the BTU value of coal (12,500) by the quantity purchased (2000 pounds–one ton) by the efficiency of a coal furnace (0.60). The answer is 15,000,000 BTUs per ton.

Now to determine the comparable heat value of other fuels, use the following equation:

BTU/unit x Quantity x Efficiency = 1
Electricity 3413 BTU/KWH x Q x 1 = 15,000,000
Answer: Q = 4365 kilowatt hours of electricity

Gas 1000 BTU/cu. ft. x Q x .75 = 15,000,00
Answer: Q = 20,000 cubic feet of gas

Oil 138,000 BTU/gal. x Q x .65 = 15,000,000
Answer: Q = 167 gallons of oil

Wood 7950 BTU/1b. x Q x .5 = 15,000,000
Answer: Q = 3773 pounds of wood = 1 to 1.2 standard cords

To summarize, one ton of coal with a heat content of twelve thousand, five hundred BTU's per pound provides approximately as much heat as:

4365 KWH of electricity
20,000 cubic feet of natural gas
167 gallons of fuel oil
1–1.2 cords of seasoned hardwood

(Note: If the coal you buy has a higher or lower BTU value than twelve thousand, five hundred, calculate its BTU value per ton and use this figure in the equation to determine the comparable heat value of other fuels.)

From here on it's easy to figure out how much it costs to heat with the different fuels on a comparable basis:

One ton of coal @ $105.00 per ton = $105.00
4365 KWH of electricity @ 7 cents per KWH = $305.55
20,000 cubic feet of gas @ 0.36 cents per cubic foot = $72.00
167 gallons of fuel oil @ $1.00 per gallon = $167.00

1.1 cords of wood @ $95.00 per cord = $104.50

Substitute the prices people are paying in your own community for the above cost figures. If the figures you come up with make coal look like a good buy, the next step you should take is to find out what it would cost to convert your present heating plant to coal or to put in a new coal furnace or stove (or two).

To do this, look up how much fuel you used last year and what it cost. Determine how much coal is the equivalent of last year's fuel and what it would cost. Subtract the cost of coal from the cost of your present fuel. Then divide the answer into the cost of installing a new coal heating plant. The answer will be the number of years it will take you to recoup the cost of converting to coal. From there on you will be saving money.

For example, let's say you have an old oil-fired warm-air system. Last year you used eight hundred and thirty-five gallons of oil costing eight hundred and thirty-five dollars. To produce the same amount of heat this year and thereafter, you'll have to buy five tons of coal costing five hundred and twenty-five dollars. This will reduce your annual fuel bill by three hundred and ten dollars. If a new coal-fired furnace costs two thousand dollars, your annual fuel saving will pay off the cost in about six and one half years. From there on, you're in the clear and coining money.

Ceramic-tile stoves are colorful and easier to keep looking clean than steel or cast-iron stoves. The grilled door is strictly for ornament. The firing door with a glass panel is behind it. (Ceramic Radiant Heat)

Chapter 3
Heating with coal stoves

Whoever thought that Twentieth Century Americans would have a love affair with stoves?

In 1976, when Stanley wrote a book about fireplaces, he decided after some debate with himself to include a chapter on stoves; and after diligent search found fifteen stove manufacturers to list. Today there are entire books about stoves and a hundred or or more firms making or importing them.

It's a good example of American unpredictability.

In the past, stoves heated many homes—mainly in the country—grocery and dry good stores (who hasn't seen pictures of countrymen sitting around a stove, chewing tobacco and swapping yarns?) and other small business establishments. Even just before OPEC became an epithet, stoves still heated vacation homes and small year-round southern homes (where you need heat one day and not the next). But today you find them everywhere. To be sure New England and the other eastern states boast most of them, because this is the land of oil heat. But you'll also find a great many in other regions that are substantially forested in homes of all sizes.

Understandably, wood is the favorite fuel. This is partly because it's traditional; mainly because it's plentiful and cheap. At least it used to be cheap. It isn't now and that may have been the first reason why coal stoves began to enter the picture. But there are others; and taken in aggregate they've convinced a good many people that they should either switch from a wood stove to a coal stove or buy a coal stove in the first place. Seeing the handwriting on the wall, a sizable number of wood-stove manufacturers are now selling coal stoves, too.

Why heat with a stove?

If you have never heated your home or small business place with a stove, the first question you have to answer is: "Should I?"

We'll tell you why you shouldn't first.

1. It's because running a stove, whether it burns coal or wood, involves a certain amount of work. You have to store the fuel (that also takes space). You have to bring the fuel in in small loads and store it again for when it's needed. You have to build the fire and tend it. You have to remove and dispose of the ashes. All this takes a little time and some muscular effort. If you're a confirmed user of a fireplace, the work is familiar and no more or less arduous. But if you've never done any more than flick a thermostat to turn up the heat in your house, it's a whole new way of life.

2. A stove may not seem appropriate to your decor. To be fully effective as a heater, it should stand out from a wall, and you may not like its looks. However, there are stoves that can be set into a fireplace, and there are others of such beauty that they are quite suitable in an elegant parlor.

3. Most—but not all—stoves get so hot that when you touch them they give you a nasty burn.

4. They do not heat an area as uniformly as other heating devices. It's always hotter close to the stove than it is farther away.

So much for the drawbacks. The pluses are these:

1. If it's an efficient design, a stove is no more expensive to run than a central heating plant. It heats the room around it by radiation and convection and the adjacent rooms by convection and to a certain extent by conduction. By contrast, a warm-air heating system heats largely by convection and to a slight extent by conduction. Hot-water and steam systems heat by radiation, convection and to a slight extent conduction. Only electric room heaters, which supply heat in various ways, heat more efficiently than a stove; but electric heat is expensive (unless you live in a warm climate and use a heat pump).

2. The total installation cost of a stove is considerably less than that of any other type of heating system. In fact, you can usually install two stoves (as many people do) for less than some other form of heating.

3. Most home owners use stoves to supplement a central heating system. That is, they use the stove to heat the house on

mild days or up to a tolerable temperature on cold days and turn on the central plant only when additional heat is required. Or they use the stove to heat the kitchen, living room, family room area twenty-four hours a day; the central plant to warm the bedrooms only when they get up and go to bed (or they may not heat the bedrooms at all).

4. Many owners of small, compact houses, on the other hand, use a stove—or perhaps two—to heat the entire house.

5. Actually, the ways in which people heat with stoves are legion. They follow no set pattern. They may try one method, then change to another and another until they hit on the one that is most satisfactory. This is one of the prime advantages of stoves. As long as you can easily and cheaply arrange for the venting of a stove, you can use it as a semi-permanent space heater.

Coal vs. wood stoves

If you can cut your own wood or have access to a steady, low-cost supply, you can't beat a wood stove. But if you tire of cutting and splitting wood or your supplier jacks up his prices to the current market price of close to one hundred dollars a cord, look into coal.

Coal has considerably more heat value, so even if a ton costs more than a cord of wood, it's still to your economic advantage to use it. At the prices given in Chapter 2, one ton of coal is the equivalent of 1.1 cords of the very best hardwood and two cords of softwood.

A coal fire is easier to tend than a wood fire because it requires less attention. You usually have to load it only twice a day—in the morning and evening. Between times it burns slowly and consistently, and you need not keep peeking at it to make sure it's going. You do have to remove the ashes more frequently than from a wood stove.

People who have burned both coal and wood say that coal is less messy. You'll probably have to wash your hands after handling it, but it doesn't leave behind splinters and bits of bark as you bring it in and lift out pieces for the fire. You can carry it in a bucket and store it by the stove in the same bucket—which means that it takes up less space near the stove (wood takes up a lot of space, which is not always available).

Even if you buy coal in quantity, it takes up a third or more less storage space than the equivalent quantity of wood. Of course it does have to be stored indoors, while you can keep wood outdoors.

When you bring coal into the house, you don't bring termites and other pests with it. Wood stove owners don't think of this very often, if at all, but they do run the risk of infesting their homes.

Another major benefit of coal is that it doesn't cake up the stove pipe and chimney with creosote as wood does. It does leave some soot and fly ash, but this is easily cleaned out and isn't dangerous if it isn't. Creosote is extremely dangerous and is the cause of many of the fires that have been blamed on wood stoves.

The main drawbacks of coal are the volatiles it contains.

In areas where wood stoves are widely used—Vermont and the skiing areas of Colorado, for example—they were this year widely blamed for the pollution of the atmosphere. Unfortunately, coal—much of it anyway—contains more pollutants than wood.

Coal is also a worse source of carbon monoxide than wood, but you don't have to worry about this if you manage your stove properly.

When fresh coal is added to a stove, it may—depending on the type—give off the odor of sewer gas until it is burning briskly. Wood smoke is acrid but not quite so unpleasant as coal smoke.

Types of coal stove

Coal stoves are designed for use with all types of coal except cannel coal. Some can be used with wood, too. More about this later.

Since coal is just beginning to catch on as a stove fuel, there are not so many coal stoves on the market as wood. But they are generally comparable to wood stoves.

To begin with, they're made of cast iron, steel or ceramic tile.

Cast iron has always been thought to be better than steel because of its greater weight, because it holds heat longer after the fire has died down and because it lasts longer. It may crack but rarely does.

But perhaps because steel costs the manufacturer less than cast iron and he can therefore sell stoves at a little lower price, current opinion holds that steel is just about as good as cast iron. It heats up faster though it cools down faster, too. It may warp if continually overheated, but this does not affect its heating ability. However, this applies only to steel of a thickness comparable to cast iron. If you buy a very cheap stove made of thin sheet steel, it probably won't last more than a year.

Both metals rust if you don't watch out, so they should be kept away from moisture and need to be blackened with stove blacking now and then even so. Both metals should, as a rule, be lined with firebrick when used with coal because coal burns hotter than wood. (However, once you've installed the firebrick—some stoves come with it already in place—you don't have to worry about the bricks. One may disintegrate and need replacement, but this is rare.)

Ceramic tile stoves come from Europe and are more expensive than metal stoves. Generally, they are more attractive because of their design and color. You can put one in the front parlor and never apologize for it or have to fret about its discoloring or rusting.

Functionally, while a ceramic tile stove may seem to give off less heat than a metal stove, this is illusory. It simply gives a lower level of heat; you won't necessarily be burned by it if you touch it. Because of tile's low thermal conductivity, a stove takes a while to emit warmth when started, but it radiates warmth long after a steel stove has cooled down and for some time after a cast iron stove.

Three basic types of stove are available:

Circulating stoves are unique in that, if they produce any heat by radiation, it is minimal. They look like over-size metal boxes and there's nothing attractive about them. But they do an excellent heating job and until modern airtight radiant stoves appeared, they were used almost exclusively for heating whole houses. They're designed to produce a lot of heat, and they do this gently and evenly. A person sitting close to a circulating stove isn't much warmer than one sitting across the room. This is because the outer shell or jacket is relatively cool—cool enough to touch.

The stoves have two walls. The inner wall, enclosing the firebox, is separated from the outer wall, or jacket, by an air space several inches wide. The outer wall contains louvers or perforations. Heat from the fire passes through the inner wall by conduction and heats the air circulating through the air space. The warm air then circulates through the room in a doughnut-shaped pattern. As it leaves the stove, it is replaced by cold air pulled in from the floor level. Thus air circulates continuously through the room and adjoining rooms.

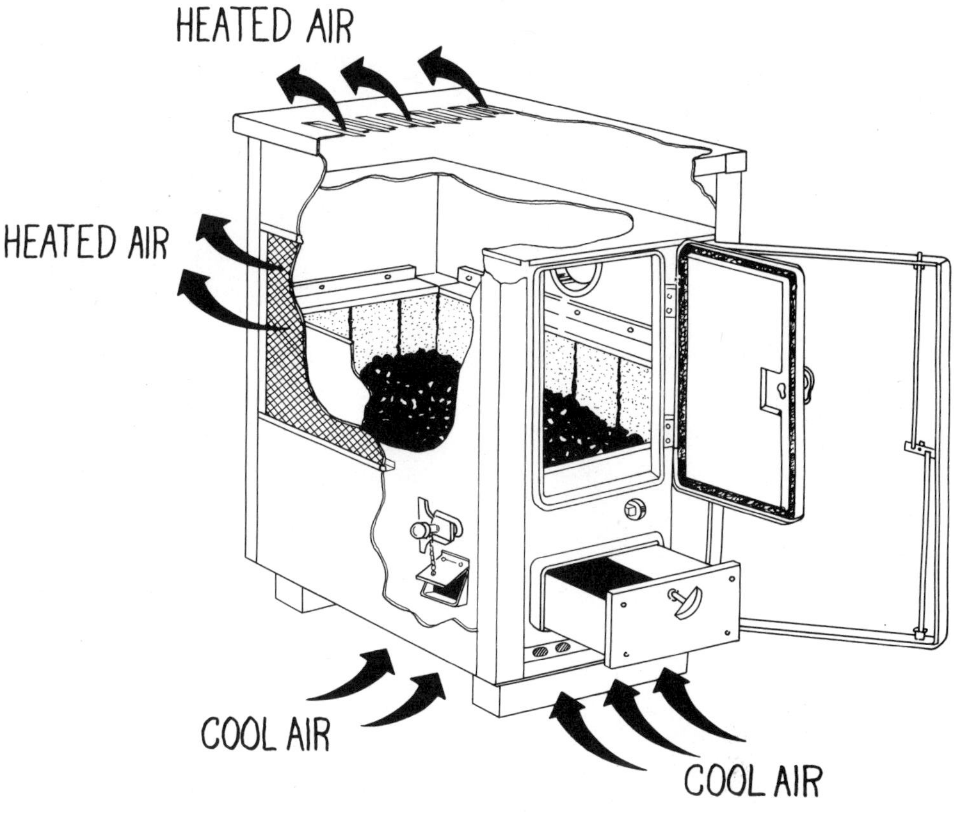

A circulating stove is a double-walled unit. Cool air introduced between the walls is heated and flows out into the room. A fan can be added to increase air flow.

← *Circulating stoves are not very attractive but are highly efficient. Because the outer shell is not exposed directly to the fire, they can be set closer to combustible surfaces than other stoves. (Suburban)*

Radiant stoves. These are the stoves the uninitiated usually think of first. There are two classifications: Those known as box stoves, parlor stoves and potbelly stoves, and the Franklin stoves.

The box stoves, etc., are cheap, inefficient and mainly small. Although they have draft controls, the joints between seams and around the doors are so loose that unwanted air seeps constantly into the firebox, thus hastening the consumption of wood and even encouraging it to burn out of control. This isn't to say that a box stove doesn't radiate a lot of heat; but the fire burns out pretty quickly unless it is frequently replenished. Result: You use more coal than you have to.

Franklin stoves, named after old Ben, are charming to look at and you can use them as a fireplace simply by opening the large doors—thus increasing their charm. But the doors are so loose fitting that, when closed, a great deal of air is drawn into the firebox through them; consequently, like the box stoves, they use up fuel at a rapid rate and don't carry through the night.

1. Superior.

2. Classic Stove.

Coal and coal/wood stoves are not available in such variety as wood stoves but the number is increasing rapidly. Here's a representative collection. Glass fronts are fairly common. 1. Superior. 2. Classic Stove. 3. All Nighter. 4. Crane. 5. Bow & Arrow. 6. Fisher.

3. *All Nighter.*

4. *Cran*

5. *Bow & Arrow.*

6. *Fisher.*

Airtight radiant stoves are far preferable to the old, loose-jointed designs—and are far more widely used today—because you have almost perfect control of the fire. Very, very little air seeps into them around the doors and through the joints (but there is enough so that, even if you close the damper and air intake, the fire won't go out right away); consequently the fuel burns quietly (unless you open the draft controls) and lasts a long time. Once you get the hang of operating a stove, you need to load it only twice a day to get heat around the clock. Since there is also little fluctuation in the fire, the heat given off by the stove is more uniform.

The principal criticism of airtight stoves is that, if you try to make a load of wood last too long by nursing it along at a very slow-burning rate, the volatile gasses given off are not burned within the stove; the flue temperature drops; and creosote builds up faster than it should on the cold walls of the flue. But since coal does not form creosote, you need not worry about this. On the other hand, if you run a coal fire at a low temperature, the carbon monoxide level rises and may leak into the house.

Variations of the basic stoves

While all stoves fit into one of the three basic categories, they vary in minor ways. This is particularly true of the airtight stoves.

Thermostatic controls. These are intended to ensure that, as long as the fire is burning, your room temperature will automatically hold more or less steady. The thermostats are similar to those on central heating systems: Turn them down and the fire burns lower; turn them up, it burns hotter. If set in one position, the thermostats continuously measure the temperature of the room air and that of the stove surface and adjust the amount of air entering the stove accordingly.

Primary and secondary air inlets. The main idea here is to extract the maximum heat from the coal by burning the volatiles that would ordinarily escape up the chimney. This is accomplished by the secondary air inlet. The primary inlet provides the air that helps to start the fire and keep it perking.

Baffles. Baffles within the firebox slow the air moving through it. This has two immediate effects: First the stove's radiation surface is enlarged and therefore throws off more heat. Second, the volatile matter cannot escape so rapidly, so more of it burns, thus increasing fuel efficiency and heat output.

Air circulators. These are, in effect, an appendage on a stove, designed to augment heat output. A circulator is a small, thermostatically controlled electric fan that passes cool room air through the stove, where it is heated by exposure to metal plates (not by direct exposure to the fire) and blows on out into the room.

A fan at the right front of this stove blows heated air into the room, adding to the heat radiated from the stove body. (Old Erie)

Fireplace inserts. One of the reasons Stanley hesitated about installing a stove was that he and Elizabeth didn't think it would look well sitting out in the middle of the room, so to speak. Eventually they found a stove that could be set into the fireplace where it looked more or less natural and the unattractive stovepipe could be hidden. But as a heat source it failed pretty badly because it radiated heat into the room only from the front. Most of the heat from the sides, back and top of the stove went up the chimney.

To solve this problem, some companies now make fireplace-insert stoves that are specifically designed for installing "in" fireplaces. The type that actually stands in front of the fireplace is as efficient as a conventional stove. Those that are recessed in the firebox are not because too much of the heat radiates toward the fireplace walls; but they are better than an ordinary stove that just happens to fit into the fireplace. The reason for this is that they are attached to and surrounded by a metal plate that completely covers the fireplace opening. The plate increases the stove's radiation surface besides preventing loss of heated room air up the chimney.

Fireplace-insert stoves are designed to take up less floor space without as much loss of efficiency as you get if you push a conventional stove into a fireplace. The metal shield closes the fireplace opening to prevent heat loss up the chimney and also picks up some heat from the stove and radiates it into the room. (Old Erie)

GLACIER BAY

Cookstoves. A number of conventional airtight and non-airtight stoves have lift-off circular plates in the top so you can cook on them. But the cookstoves we're talking about are honest-to-goodness kitchen ranges with several surface "burners," an oven and a firebox. You've seen pictures of them in old-time kitchens and sometimes you can pick up one of these antiques; but there are also modern, enamel-covered stoves of pretty good airtight design. In some of these you burn coal (or wood) alone. Others have a coal (or wood) firebox on one side and a second firebox for gas or oil. There are also electric ranges with built-in coal heaters.

It seems hardly likely that you are enough of an enthusiast to want to return to the days of cooking all meals over coal. But if you have a large enough kitchen and live in the country, you might want a cookstove–in addition to an everyday electric stove–to heat the kitchen and to use for cooking when the power fails.

The entire top of this coal-fired cookstove is a giant surface burner. The oven is in the center. Coal is loaded through the upper, left door and ashes are removed through the lower door. Below the white lift-off panel in the top is a hot water tank. (Merry Music Box)

STYRIA

Coal Heat

Stove cost

You can probably find some box and potbelly stoves in the one hundred dollar range; but the saving you make initially will be eaten up rapidly by the large amount of coal you use. Efficient fuel-conserving stoves start at around three hundred dollars and range upward to about eight hundred dollars. If you fall for a ceramic-tile stove, don't be surprised if it costs over one thousand dollars.

Cost, obviously, is a major consideration in buying anything; but don't let it rule you too much when it comes to buying a coal stove. Efficiency, heat output and durability are much more important in the long run.

Sizing the stove

The heating capacity of a stove is given by the manufacturer in cubic feet and/or BTUs. The BTU figure is better to work with. The largest stove we've found has a thirty thousand BTU output.

Before shopping for a stove, you should decide how much of the house you want it to heat. One room? Two or three connecting rooms? The whole house? You must remember that if you want to heat more than one room, the fewer the walls there are, the better heat will flow from the stove room into the others. This is not to say that if two rooms are connected only by a conventional thirty inch wide door, the heat from the stove won't eventually flow into the next room through the door and by conduction through the wall; but the process will be fairly slow unless you install a fan high up in the dividing wall to hasten the transfer. (Fans are made especially for this purpose, although you can use an ordinary kitchen or bathroom exhaust fan).

Heat can also be transferred from one story to another, which is why a good many people install stoves in a basement. As the basement warms up, it warms the floor above and this in turn warms the rooms above. But here again the process is slow unless you install grilled openings in the floor, leave the basement door open or actually pump the heat upstairs. Cary and her husband use the pumping arrangement with almost startling results. Charles put a chimney heat reclaimer in the stovepipe above his basement stove, and connected it by a duct to a floor outlet in the living room. Together with the heat conducted from the basement through the living room floor, the heat reclaimer puts out enough BTUs to bring the living room up to seventy degrees. But we mention this only in passing. When you are determining the stove size you need, don't figure on using a heat reclaimer. The BTU output of the stove alone is your one and only concern.

Once you have settled on the space to be heated, measure the length and width of the floor area and multiply them to find the square footage. Then multiply this figure by the ceiling height to give you the cubic footage.

Now you run into a real problem: Calculating the size of any heating plant is a very complicated job, and we don't advise that you try it. If you're heating a new building, you should, instead, hire a heating engineer to make the calculation for you. If you're heating an existing building, measure the heated area of the entire house and calculate its cubage. Find the BTU capacity of your present heating plant on the nameplate. Then divide the cubage of the house into the BTU capacity of the furnace and multiply the answer by the cubage of the area you want to heat with a stove. The answer won't be one hundred percent accurate but it will be close enough. (If the furnace nameplate doesn't mention BTU capacity, you should either employ a heating engineer to determine what stove you need or go by the cubic-foot, or sometimes square-foot, rating assigned to the stove by the manufacturer. The latter method is only semi-accurate, but often proves satisfactory. If it's not, you can sell the stove and get a better one. The market for used stoves is excellent.)

One final point: Once you've hit on the stove size you need by one of these methods of calculation, don't arbitrarily conclude that you should buy a larger model. It will put out so much heat that you'll be driven out of the room in which it's located.

Coal Heat

Buying a stove

In addition to buying a stove of the correct size, there are several things to consider when shopping for one. (Incidentally, you may have to devote several days to your shopping expedition because no stove manufacturer has such wide distribution as automobile and appliance manufacturers have and because few stove dealers carry more than three or four brands.)

1. Make sure the stove is well made; otherwise you may have to replace it soon because it falls apart or proves unsafe. The walls should be pretty thick and free of cracks or open seams (stick your face into the loading door and see whether any lines or spots of light show through the shell). The door should close tight; when you close it over a dollar bill, you shouldn't be able to pull out the bill without tearing it. The grate and liner should be heavy and well constructed.

The extent of the manufacturer's warranty will give you a good idea whether he thinks well of the stove.

2. Do not buy a stove unless it has been tested and approved by Underwriters Laboratories (UL) or a named independent testing lab. In some states today you are required to buy a stove that is listed as having been tested by a recognized laboratory. If you buy an unlisted stove, the installation may not be approved by your local building inspector (more and more states and cities insist that all new stove installations be inspected and approved). And if an unlisted stove should cause a fire, you might not be fully covered by your insurance.

At the moment, unfortunately, United States and Canadian testing labs do not concern themselves with stove performance (as British labs do). They want to know only whether a stove is safe; and to this end, they subject all stoves that are submitted to them to grueling tests. First they test to find out whether the stove can be safely used with the fuel recommended by the manufacturer. Then, on the theory that consumers may not always use recommended fuels, they make tests with other fuels. The stove is not listed as approved unless it proves safe with all fuels (but it should be used only with the fuel recommended by the maker).

Look for the testing laboratory's label on the stove.

3. Make sure the collar to which the stovepipe is connected is located so you can install the stove without excessive reconstruction of the house. On some stoves the pipe connects to the top of the stove; on some it connects to the back or side. Others make provision so you can connect the pipe either to the top, back or side.

What makes a good coal stove

The things to consider in buying a stove do not stop with good workmanship and a testing laboratory label. If you want an efficient stove—as of course you do—you have to know what distinguishes these from run-of-the-mill stoves.

1. A stove specifically designed to burn coal is more efficient than one that burns either coal or wood. (A stove designed for wood alone should never be used to burn coal because it isn't made to withstand coal's higher heat output.) One of the top stove makers puts this very clearly and honestly in his description of his wood/coal stoves. "The use of the coal grate," he says, "is basically a compromise. No combination-fuel stove is as effective as one designed specifically for one fuel only."

2. A coal stove has a slightly larger draft opening than a wood stove because coal requires about twice as much air as wood to raise it to the temperature where it starts to change from its solid state to a combustible gas.

Generally, the draft opening in a stove meant for bituminous coal is larger than one in an anthracite stove. Bituminous stoves need a lot of air to burn off the volatiles quickly. An anthracite stove uses less air to prevent overheating of the stove.

3. In a coal stove, the air comes in under the grate so that it can pass upward through the bed of fuel; otherwise the fire dies out for lack of oxygen.

4. A coal stove must have a heavy (usually cast iron) grate with

openings of the proper width to hold the type of coal specified for use. Provision should also be made for shaking the ashes out of the grate into the bottom of the stove.

5. An ash pan for easy removal of the ashes is included in the stove. This simplifies the daily removal that is necessary.

6. In an anthracite-burning stove, the firebox is smaller than that in a wood stove. This is to assure that the coal in the back of the stove will be burned. The firebox is also higher to permit building up a sufficient thickness of coal to give a longer burn period. It should be high enough so you can have a fire bed eight to ten inches thick. Such thickness is possible since anthracite does not fuse together as it burns.

In a bituminous stove, the size of the firebox is less important because the coal burns more like wood. However, the firebox cannot be too large; otherwise it might be loaded with as much as one hundred pounds of coal at a time. For best combustion, loading of forty to fifty pounds is generally recommended.

Note: A firepot is the same as a firebox. The word is used when the stove is round rather than rectangular.

7. Coal stoves generally are lined with cast iron or, more often, a refractory material such as firebrick to protect the stove walls from the high heat. This does not mean, however, that a steel-walled stove that passes laboratory tests may not be approved for use. Remember that United States and Canadian laboratories are concerned only with safety, not performance or durability. In other words, if a steel-walled stove is approved, it is only on the grounds of safety.

8. A stovepipe damper is unnecessary for an airtight stove since you can control the air supply adequately by the air inlet on the stove itself. With a non-airtight stove, however, a stovepipe damper is desirable to prevent the fire from burning too briskly.

If a stovepipe damper is recommended by the stove manufacturer, it is generally thought that one with a small hole in the center is better than a solid damper since it keeps you from closing off the flue completely, thus preventing the escape of carbon monoxide.

Whether a coal stove should have thermostatic controls, a secondary air intake or baffles is largely a matter of personal preference. Good stoves are made with and without them.

There is no doubt that a thermostat is a convenience. Baffles probably do help to extract extra BTUs from the fuel, but they have less effect in coal stoves than in wood stoves because coal fireboxes are smaller. In a bituminous stove, a secondary air intake ensures admission of ample air to burn off the volatiles quickly, thus reducing smoke and carbon monoxide danger and increasing the coal's heat output. But it is less important in an anthracite stove.

Converting wood stoves to coal

Inevitably, some wood-stove owners who decide they would rather burn coal wonder whether they can use their present stove instead of buying a new one. And if so, what if anything has to be done to the stove?

The answer to the first question is "maybe." If the manufacturer's literature clearly states that your stove can be used with wood or coal, you can obviously make the change. On the other hand, if the manufacturer's literature and/or the testing laboratory's label says the stove can be used only with "solid fuels," you're in a quandary. Theoretically "solid fuels" covers both wood and coal; but perhaps in the case of your stove it means only wood. To clarify the situation, you must either talk to the stove dealer or manufacturer.

Having settled this point—assuming that your wood stove can be used with coal—you should consider the consequences of making the conversion.

We mentioned one of them already: No combination-fuel stove is as efficient as one designed for a single fuel.

The second point to note is that converting may or may not be easy. Read the manufacturer's instructions or consult your dealer. It may be that all you have to do is install a coal grate and ash pan. On the other hand, additional work may be required. Here's what one manufacturer has to say on the subject: "To use the coal grate you must remove the air baffle from the back of the stove door. This is to allow the gasket on the front edge of the grate to fit snugly against the door. This way the air entering through the draft caps is directed under the grate and up through the coal bed to supply the oxygen for the burning coal. When you go back to burning wood, you should replace the air baffle on the door. One of the most common problems associated with the use of this coal grate comes not from the grate but from the stove owners. They have usually been burning wood for a season or two. Being used to using their stoves with wood for so long, they tend to set the draft caps on the doors in the same position as they did with wood. This promptly suffocates the fire, causing it to go out."

Installing a coal stove

The installation of a coal stove is no different from the installation of a wood stove. You must provide a safe location for the stove and a flue to carry off the products of combustion. Be sure to read and follow your local building code's requirements.

Your first thoughts should be directed to the safety of the installation. If you get right down to base facts, all heating devices are dangerous; but a stove—whether it burns coal or wood—is especially so because it is most often placed right in the living area, emits a great deal of heat and—since it is fired by hand, often by everyone in the home—is often misused or actually abused.

Here, then, are the rules for installing it so it will be safe:

If the stove is of the radiant type, it must be placed at least thirty-six inches from all combustible surfaces, except the floor. The clearances are measured from the sides, back, front and top of the outside surfaces of the stove.

Note: The National Fire Protection Association has found that almost all wall and ceiling surfaces except those made of solid masonry are combustible. This includes surfaces that are covered with sheet metal, brick and stone veneer, plaster and gypsum board.

Unless it's properly installed, a stove can be very dangerous. Adequate clearance from combustibles is mandatory. Covering a wall with a four inch thickness of brick with a one inch air space behind it permits a reduced clearance and is very decorative as well. The floor must also be protected. The barometric draft regulator on the stovepipe here is unusual but sometimes recommended. (Heath-delle Associates)

Even solid masonry is combustible if covered with paint.

Radiant stove clearances can be reduced, however, if a combustible surface is protected in the following manner:

	Clearance in Inches		
	At sides and back	In front	At top
¼" asbestos-cement board over 1" air space	18	24	30
28-gauge sheet metal on ¼" asbestos-cement board	18	24	24
28-gauge sheet metal over 1" air space	12	24	18
28-gauge sheet metal on 1/8" asbestos-cement board over 1" air space	12	24	18
4" brick over 1" air space	12	24	
22-gauge sheet metal on 1" mineral-wool batts reinforced with wire mesh	12	24	18

The protected surface should extend at least eighteen inches beyond the sides and top of the stove. To provide the one inch air gap behind the protective sheet, do not use furring strips. Instead you should use one inch lengths of metal pipe spaced sixteen inches in all directions. Drive long screws through the protector and pipes into the studs. This permits free air circulation in all directions and there is no wood that can be charred by the heat.

The clearances between a circulating stove and combustible

flagstone or ceramic tile. But we can not vouch for the effectiveness of these materials or the attitude of building inspectors toward them.

If a stove is installed in or in front of a fireplace, the hearth may have to be extended with a floor protector, and the clearances called for above are required between the sides and top of the stove and the mantel or any other combustible material on the fireplace breast. The alternative is to install between the mantel and other combustible materials and the stove a heat shield of metal or asbestos-cement board.

The actual installation of a coal stove, once you've made the necessary provisions to protect the floor and nearby walls, involves nothing more than assembling whatever parts the dealer has left unassembled, setting the stove in place and connecting it to the chimney.

Connecting stove to a masonry chimney

If you have a masonry chimney you can connect the stove to, you're ahead of the game.

Of course, not all chimneys are perfect. An inside chimney is better than one on an outside wall because it doesn't get so cold (resulting in a slow starting draft) and isn't so likely to draw in outside air while the fire is going. An inside chimney also gives off considerable heat to the surrounding house whereas an outside chimney does not.

Another disadvantage of both exterior and interior chimneys is that the flues are usually larger than most stoves require; consequently, the smoke from the fire is cooler than it should be and poor draft results. True, this can be corrected by dropping a stainless steel or enamel-coated steel liner (fluepipe) down the chimney to the point where connection to the stove is made. But this is easy only in a straight chimney. If it has bends (you can tell whether it does by looking up it), forget it.

The chimney may not, of course, be sound, in which case, it should not be used unless you can correct all defects—which is difficult because you probably can't find them all—or insert a new liner. You should also beware of chimneys in old houses that were used in the past for venting stoves. If the openings made in such a chimney for the stovepipes are not plugged with concrete and brick, the heat trying to escape through them may start a fire. The metal "pieplates" often used to seal stovepipe openings do not give adequate protection.

Finally, the chimney you want to connect into may already be serving the furnace, water heater, fireplace or another stove. This is perfectly permissible if the chimney has several flues and you hitch into one of the flues that is not in use. But if you hitch into a flue that is in use, there are problems:

1. Many building codes forbid connecting a stove (or other heating appliance) to a flue serving a fireplace on the theory that sparks from the former may fly out the fireplace opening into the house.

2. If, heaven help you, you should have a chimney fire, you must shut off the air flowing into the jointly used flue not only at the stove but also at the other heating appliance. It isn't easy to remember this in the emergency; and even if you do, you may have trouble stifling the air flow.

3. The flue must be large enough to supply draft for both the stove and the other heating appliance. (As a rule of thumb, the cross-sectional area of the flue must be equal to the cross-sectional area of the two vent pipes entering it.) If it's not, one or the other or both heating devices will not be adequately vented.

Related to this problem is the fact that soot from the stove may gradually choke the flue. This will reduce the stove's combustion rate but not cause serious problems with it otherwise. But if the other appliance burns gas or oil, which is injected at a steady rate, the reduced flue size may cause the unburned gases to leak out of this appliance into the house.

If despite these problems you still want to vent a coal stove through a chimney flue serving another heating unit, here's what you should do:

Connect the stovepipe from the stove (this is properly known as the stovepipe connector) into the flue several feet above or below the other stovepipe. If the latter comes from an oil or gas-burning appliance, the former should be below it.

Make sure the flue is big enough to serve both heating units (see above).

Clean the flue regularly.

Installing the stovepipe connector is largely a matter of fitting together three-foot sections of single-wall steel pipe of the size specified by the stove manufacturer. (Insulated pipes for building metal chimneys can be used in a connector but are much more expensive and much larger. Furthermore, they don't emit much heat to the surrounding air—an important consideration since much of the heat generated by a fire in a stove is given off by the stovepipe connector). Use ordinary black or blue pipe. It may not last quite as long as galvanized, nickel-plated or stainless steel pipe, but it gives off considerably more heat and looks better. The National Fire Protection Association recommends that you use twenty-six gauge pipe in connectors of up to five inch diameter; twenty-four gauge pipe in connectors of six to nine inch diameter; and twenty-two gauge pipe in connectors of ten inch diameter.

The pipe sections sold to you by the stove dealer or a hardware store will be "flat." You must form them into cylinders by joining the two edges. This is a mean job and you'd better wear gloves. The best procedure is to fit the two edges together at one end of the pipe; they will then snap together for the whole length of the pipe fairly readily. Whatever fittings, such as elbows, you need are already put together.

For a wood stove, the crimped (male) ends of the pipe sections should point toward the stove so that creosote won't leak out of the joints; but this is unnecessary for a coal stove. However, after the connector is assembled, you must drive three sheet-metal screws through each joint to hold the pipes together. The connector should also be screwed to the collar on the stove.

Although you may have seen pictures of stoves with connectors extending the entire length of a church or schoolhouse, the connector you install should be as short as possible because it will be stronger and will not reduce chimney draft severely. Ten feet is a good maximum length. It should also be as straight as possible, with not more than two ninety-degree elbows or the equivalent. The horizontal run should be sloped upwards toward the chimney about

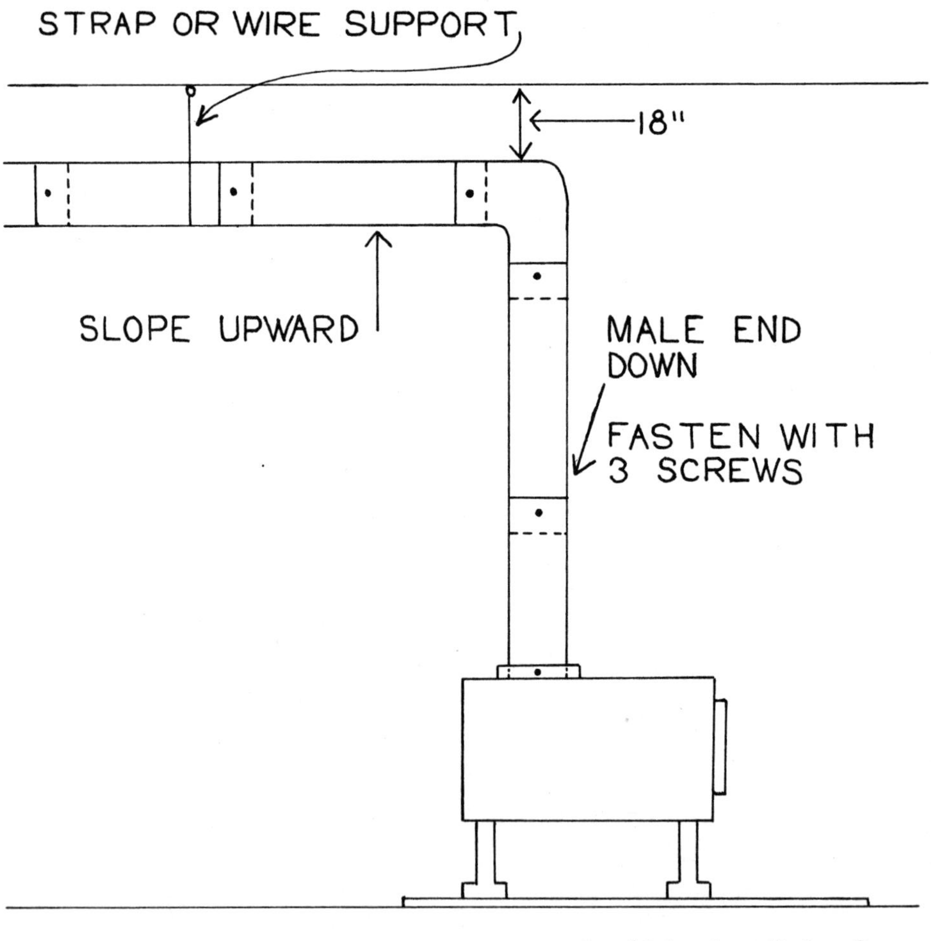

This is how a stovepipe connector should be installed. On some stoves, the pipe exits from the back or side.

¼ inch per lineal foot and should be no longer than seventy-five percent of the height of the chimney, measured from the top of the chimney to the stovepipe's point of entry. (This seventy-five percent requirement applies only to outside chimneys.) The horizontal run must also be supported at four foot intervals by steel straps or heavy wires screwed to the joists above.

If the stove is not an airtight design, a damper should be installed in the connector. Usually it is placed close to the stove but it can go anywhere.

Where a connector passes through a wall or ceiling (but you should try to avoid this if at all possible; codes generally prohibit it), substitute an insulated metal chimney section for the stovepipe section, and provide at least a two inch air space between it and the wall.

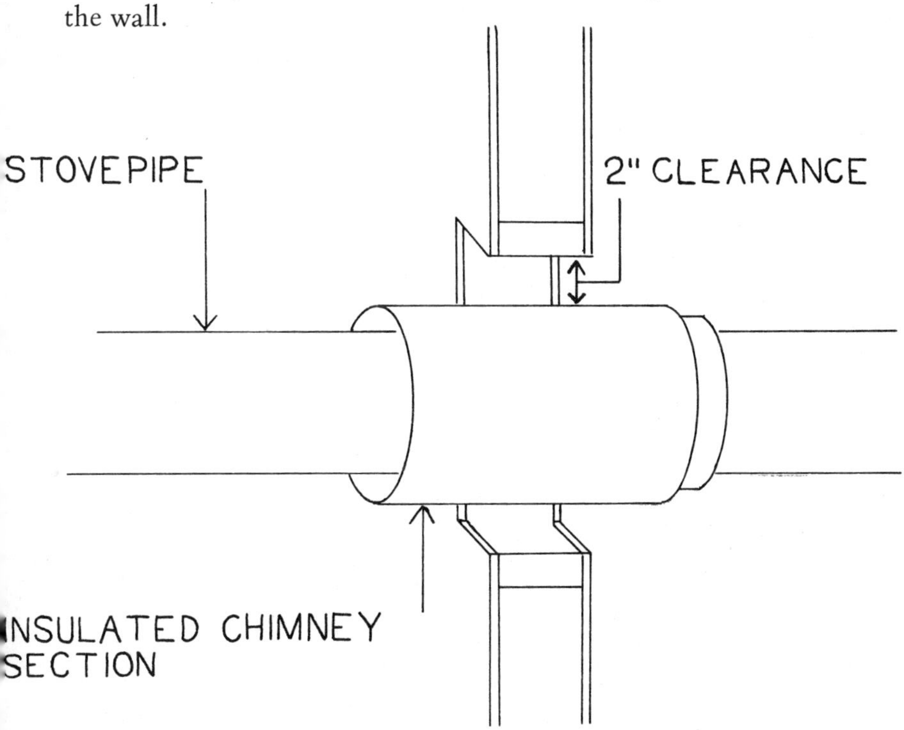

If a stovepipe must pass through a wall, substitute an insulated metal chimney section for the single-wall pipe.

The connector enters the chimney either through an existing hole (if you're lucky enough to have one) or one you make yourself with a cold chisel and masonry drill. Take special pains not to crack the flue liner outside the margins of the hole.

The hole should be just a little larger than the thimble of fireclay, 3/16 inch steel or twenty-six gauge stainless steel that goes into it. The flange on the thimble protrudes into the room, but the other end should be flush with the inside face of the chimney liner. Anchor it in the hole with mortar or furnace cement.

Insert the stovepipe in the thimble to within about two inches of the chimney liner. It does not have to be fastened so long as it cannot be pulled out of the sleeve or shoved into the liner.

If the chimney is covered with combustible material, the thimble should be long enough to extend to the surface of the combustible material. The combustible material is cut back eight inches around the entire thimble. The hole can then be covered, as in the drawing, with asbestos-cement board.

The stovepipe passes through a thimble where it enters a chimney (right). If the chimney is covered by a combustible wall, the wall around the thimble must be cut back as shown at left.

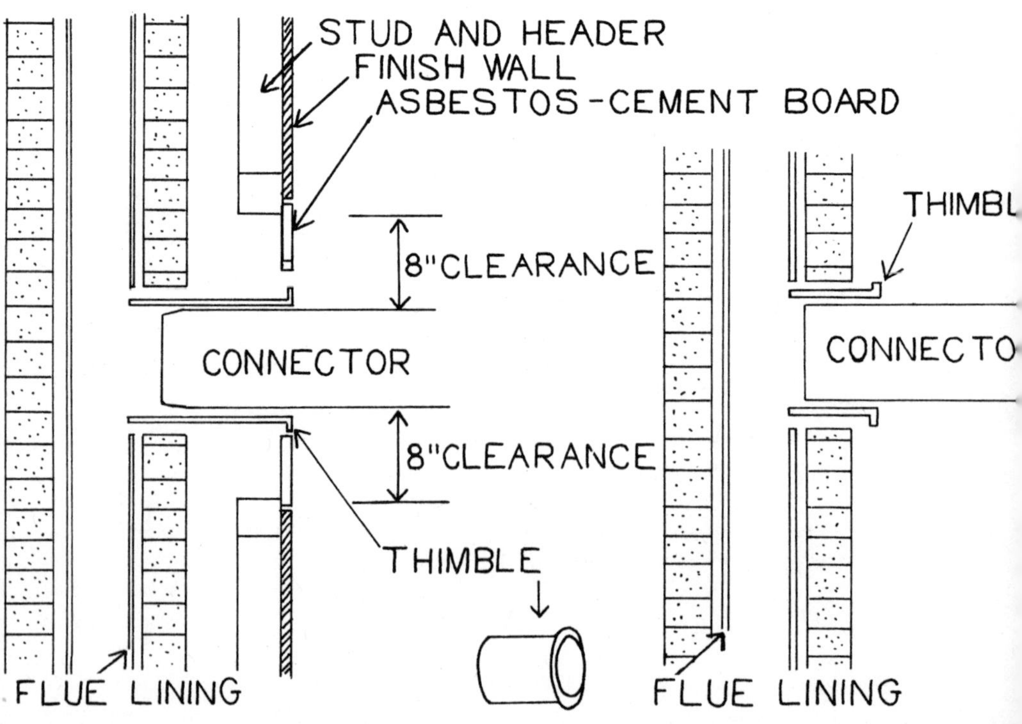

Installing a stove in a fireplace

There are several problems with fireplace installations:

1. The stove's heating effectiveness is reduced the deeper it is pushed into the fireplace. This is mainly because much of the stove's heat is radiated into the fireplace rather than into the room. It's also partly because the stovepipe is hidden in the fireplace and chimney and can't give off heat to the room.
2. The chimney flue is usually too large for the stove and results in low draft.
3. A wood mantel and other fireplace-breast trim should be shielded from the stove to prevent ignition.
4. Since a stove is heavy, installation can be difficult.

On the other hand, the stove is out of the way–you might say, where it belongs–and it doesn't upset the aesthetics and use of the room. And you don't have to invest so much money in the stovepipe.

The easiest and in most respects the best way to make an installation is to set the stove on the hearth in front of the fireplace. Three methods are shown. No. 1, with the stovepipe coming out of the top of the stove and entering the chimney above the damper, is best because the heat from the stovepipe enters the room and you don't have to dismantle the installation to clean the chimney.

The other two installations are more attractive but produce less heat and it becomes more of a chore to clean the chimney. No. 2 is the better because the projection of the stovepipe up into the chimney produces better draft. But it's a real chore to install the stovepipe and take it out and replace it during chimney cleaning. You must also seal off the flue so cold air cannot come down and cool the room and heated air can't escape. This is done by having a sheet-steel plate made to close the throat around and at both sides of the stovepipe; stuffing insulation into the throat; or covering the entire fireplace opening with a metal or asbestos-cement-board shield. (More on this later).

No. 3 installation is the easiest and cheapest. You simply cover

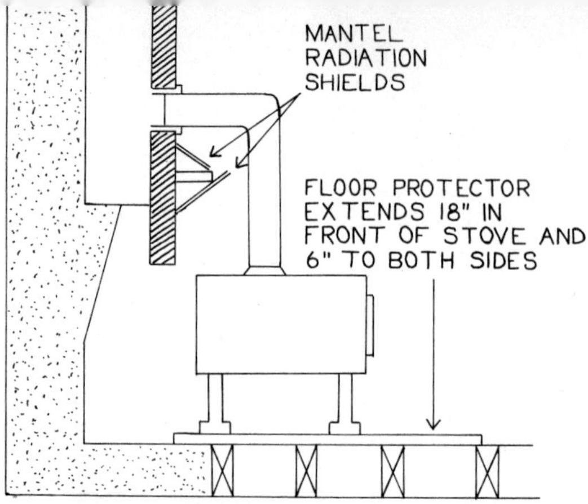

1.

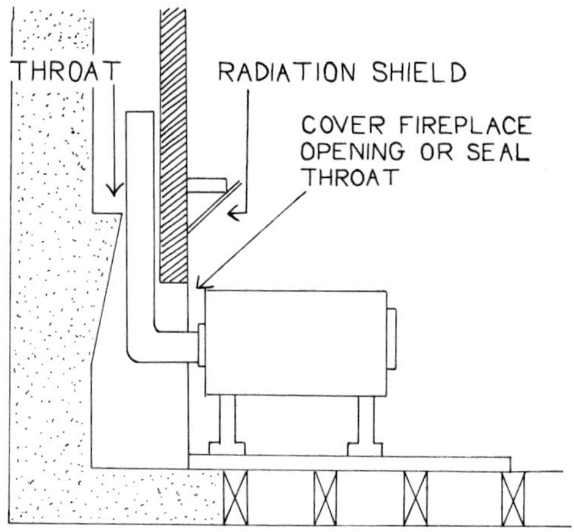

2.

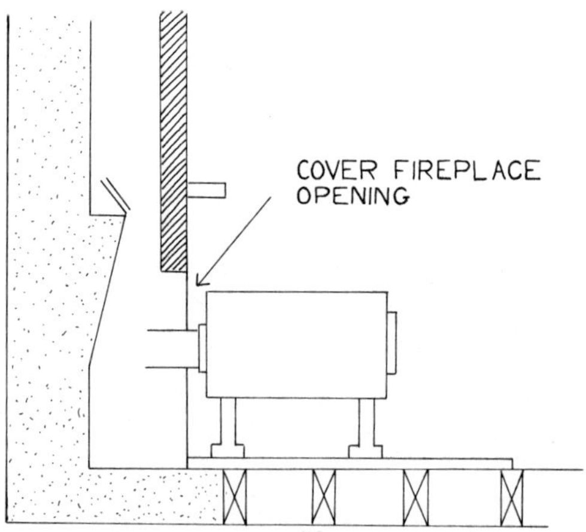

3.

Three ways to install a stove in a fireplace. No. 1 is the best because you take advantage of the heat given off by the stovepipe. No. 2 produces better draft than No. 3 but the installation is harder to make and maintain.

the fireplace opening so cold can't enter the room and heat can't escape, and insert a short straight length of stovepipe through it into the fireplace. But the draft is reduced.

If you want to stick the stove well back into the fireplace, the problems inherent in the No. 2 illustration are compounded: The heating effectiveness of the stove is reduced (and, of course, you get no heat at all from the stovepipe). The installation is a beast to make because the stove is so heavy that it's hard to position and the stovepipe is very difficult to install.

Before making such an installation, you must determine the depth (front to back) of the throat in which the damper is installed.

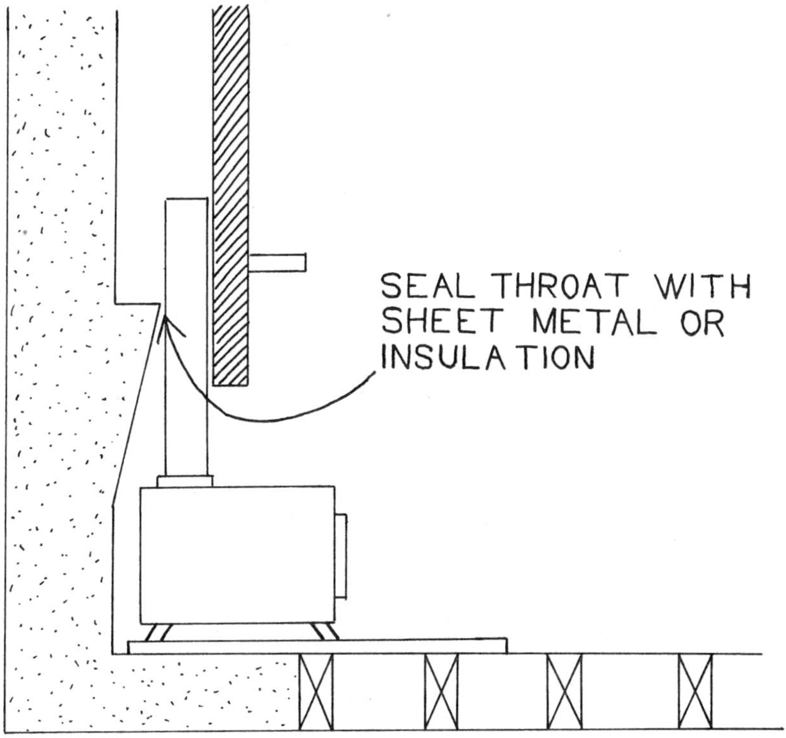

Here's a fourth way to install a stove in a fireplace. It's more attractive than the others because the stove projects less into the room, but you lose much of the heat radiating from the stove.

Even with the damper removed (it is often necessary to do this), the throat may be only a few inches deep—perhaps not deep enough to pass a stovepipe up through it. So you must either buy a stove that requires a stovepipe smaller than the throat or you must use rectangular metal ducts (like those used in warm-air heating systems) instead of standard cylindrical stovepipes. (You may be able to find these in a shop that sells warm-air heating equipment; otherwise you'll have to have them made by a sheet-metal worker.)

The stove used must, of course, have a stovepipe collar in the top. In addition, the stove must be enough lower than the fireplace opening so you can reach in and connect the stovepipe (which is first pushed up the chimney) to it. Furthermore—but this is not a requirement—the stove ought to have an air circulator so that more of the heat it generates can be pushed out into the room.

After the stove is installed, the throat must be sealed off. You can do this, as in Drawing No. 2, with sheet metal; but because it's so difficult to get into the fireplace after the stove is in place, this is a miserable job. The much easier alternative is to cram fibrous insulation into the throat. Stanley uses fiberglass without any problems, although fiberglass loses resiliency in contact with very hot surfaces. Rock wool is somewhat better. Ceramic wool, made by the Carborundum Co. and Babcock & Wilcox, is best. In any case, use insulating batts about six inches thick, and remove whatever paper backing they may have. In wind storms, the insulation is sometimes blown down into the fireplace, but it's easy enough to to stuff it back into place.

Building a new chimney

But suppose your house doesn't have a fireplace or chimney—how can you put in a coal stove?

Obviously, you build a chimney. It can be made of masonry or steel. Since the former would cost you a fortune, we'll ignore it. That leaves you with the factory-built steel chimney and it's just dandy. Much less costly than a masonry chimney, easy to install, space saving, efficient, safe. It's not pretty if you let it stick out above your house like a stovepipe; but you can surround it with a fake-brick housing (available from the chimney manufacturer) and you'll be put to it to tell it from the real thing.

For maximum efficiency, you should, if possible, install a steel chimney inside the house. If it must be outside the house, enclose it in a wood chase (long, slender box) to hide it and to increase its efficiency slightly.

When this house was remodeled, a fireplace with a prefabricated metal chimney was added on the second floor. The fireplace is enclosed in the box on the wall; the chimney is a wood chase.

The cross-sectional area of the flue you require may be specified by the stove manufacturer; if not, it should equal the cross-sectional area of the flue collar on the stove or be no more than 25% larger. The height of the chimney depends, of course, on the height of the house. If you have a flat roof or if the chimney passes through the peak of the roof, the chimney should project three feet above the roof. If the chimney emerges lower down on a pitched roof, it should be two feet higher than any part of the roof (or adjacent building) within ten feet, measured horizontally.

Metal chimneys are put together—much like a stovepipe connector—with short sections of pipe that have two to four walls separated by an air space or layer of insulation. The fittings—elbows, Tees, etc.—are similarly made. Thus it is possible to install a metal chimney within, as a rule, two inches of a combustible surface.

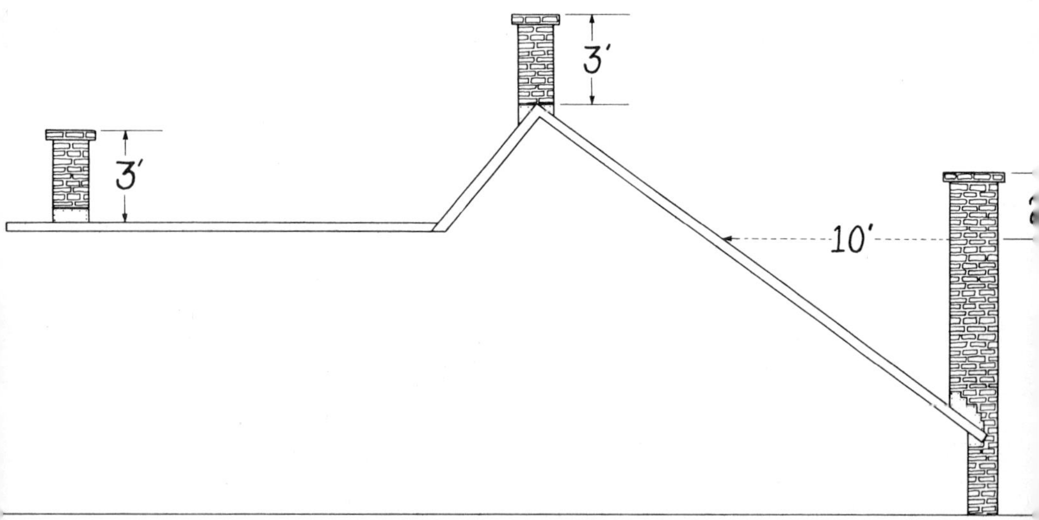

A chimney must project at least three feet above the roof or two feet higher than any part of the roof within ten feet, measured horizontally. If the chimney is within ten feet of a neighboring building or hillside, it must also be at least two feet higher than that.

The chimney you use for a coal stove should be listed by UL or other approved testing organiaztion and should be classified as a Class A, All-Fuel or Solid-Fuel chimney.

Although installation of a metal chimney can be made by any reasonably handy home owner, it is too involved to go into here. Follow the manufacturer's directions. But we do want to point out that, with a metal chimney, you have great flexibility in making an installation.

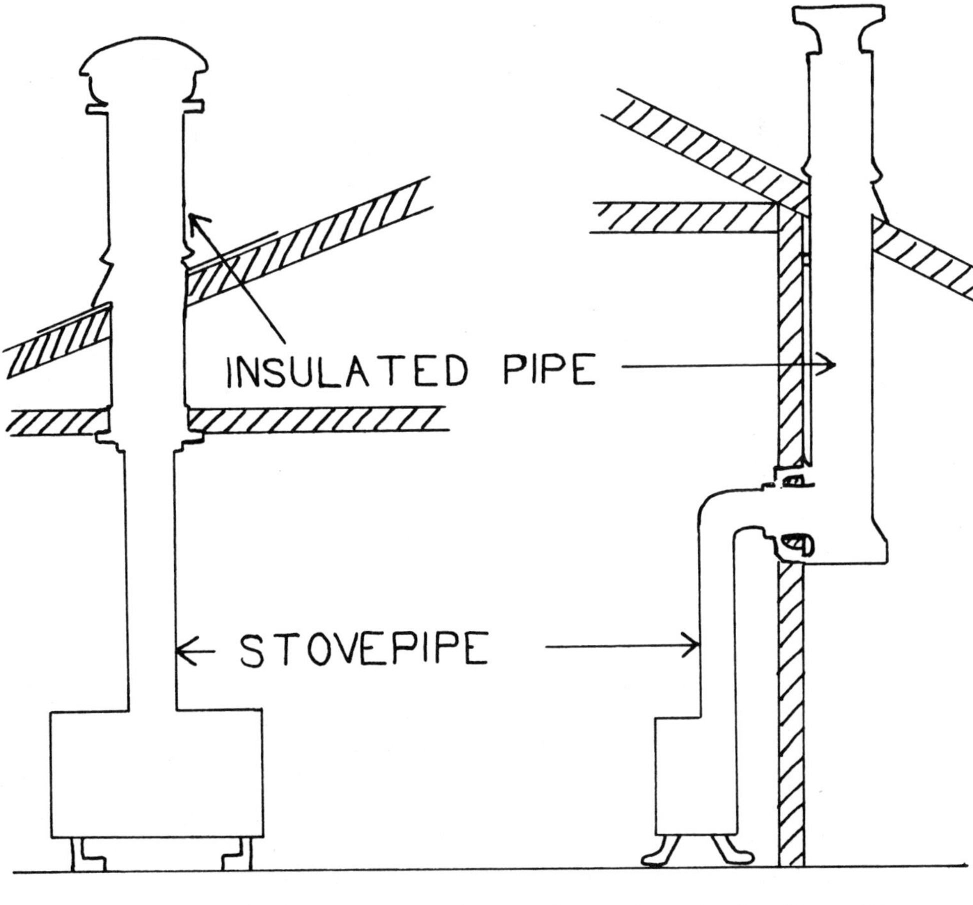

These are the two most frequently used methods of installing a prefabricated metal chimney. Others are described in the text.

A prefabricated chimney requires very little clearance from combustible surfaces. Here it's connected directly to the stove—an acceptable practice except that you get little value from the heat rising through it. A single-wall stovepipe connector is normally used.

Metal chimneys are capped to keep out rain; come with flashing collars. To improve their appearance, you can surround the extension above the roof with a fake brick housing.

The chimney is connected to the stove by a stovepipe connector. You can:

1. Run the connector straight up to the ceiling, where it joins the chimney which continues straight up through the roof.

2. Run the connector straight up to the ceiling, where it joins the chimney, which is then angled up to a point just below the roof, where it then rises straight through the roof.

3. Run the connector up to within eighteen inches of the ceiling, where it makes a ninety degree bend and continues across the ceiling to another ninety degree bend. From this point the chimney runs straight up through the roof.

4. Run the connector up to within eighteen inches of the ceiling, where it makes a ninety degree bend and continues across the ceiling to the outside wall. Here it is connected to the chimney (usually with a Tee) and the chimney runs up the wall outside the house and through the overhang.

5. Run the connector from the back of the stove horizontally to an outside wall where it is connected to a Tee. From the Tee the chimney runs up the wall outside the house.

Running a coal stove

It takes a little time to learn how to run a coal stove efficiently. So don't expect miracles for the first few days.

When starting a fire from scratch, crumple up newspaper in the grate and cover it with five or six sticks of kindling. Open the air intakes wide. When the wood is burning briskly, cover it with a thin sprinkling of coal. Then close the door and wait for about ten minutes or until the coal is burning well.

Add about three shovelfuls of coal and wait another ten minutes. Gradually build up the fire bed to about an eight inch depth. When fresh fuel is added, always leave some of the glowing coals uncovered.

Now you should have a fine bed of red-hot coals and can close the intakes. How much you close them is one of those things you'll have to learn by trial and error. If you shut off too much air, you will reduce combustion to the point where you run the risk of being asphyxiated by carbon monoxide. If you let in too much air, the fire will burn too fast and too hot.

Once you have settled on the best positions for the drafts, your fire should burn at a slow even rate at all times, and you should have to add fresh fuel only in the morning and before you go to bed.

The ashes should be shaken down out of the grate into the ash pan only about once a day. Here again you have to learn how much shaking is required. If you don't remove enough ash, the fire won't receive enough air for good combustion; if you remove too much, the fire may go out. You should shake the fire only when you have a full bed of hot coals, and then you should shake it lightly until you see a few red coals on the grate. Leave about an inch-thick layer of ash to insulate the grate from the full heat of the fire.

After shaking down the ashes, empty them into a tightly closed container. If you're feeling very pinch-penny, screen them through 3/8 inch wire mesh so you can recover whatever unburned coal has gone through the grate.

Since clinkers that may be formed will not shake through the grate, they have to be removed with tongs. They're easily distinguished from the coals because they form amorphous, rough, gray lumps.

Never poke or break up a coal fire as you do a wood fire because this brings ashes to the top of the fuel bed where they interfere with burning.

The proper sequence for firing up a coal stove when the fire has died down is as follows:

If the coals are still pretty hot—

1. Open the draft.
2. When the fire is burning brightly—about ten to fifteen minutes—shake out the ash.
3. Add coal to an eight inch level, leaving a red spot of glowing coal visible. The best procedure is to draw the topmost red coals to the front of the grate and pile the fresh coal behind them.

4. Wait ten to fifteen minutes until you have a large bed of red-hot coals and close the draft.

If the coals are nearly extinguished–

1. Open the draft.
2. When the fire begins to perk, sprinkle on just enough coal to cover the glowing coals lightly.
3. Wait ten minutes or so and add a little more coal.
4. Repeat the third step.
5. When the new coal is going strong, shake the grate lightly and repeat the third step again.
6. When you have a large bed of red-hot coals, close the drafts.

If you study these procedures closely, it will be apparent that the secret in running a coal fire is not simply to pile on more coal. First, using the drafts, get the existing coals going well. Then add coal. The fresh coal will damp the fire temporarily, but it will soon be burning briskly.

If you notice a lot of smoke issuing from your chimney, it's because the fire is not burning off the volatiles as it should. When anthracite burns with a blue flame, it's burning properly and consuming the volatiles. Soft coals burn with a yellow flame in the same circumstances. Correcting the air supply is usually the first thing you must do to improve combustion.

At night, before you retire, bank the fire so it will carry over till morning. This is done by piling the coal higher at the back of the firebox and sloping it down from there toward the loading door. Some red or burning coals must be left uncovered at the front of the firebox.

Maintaining a stove

Stoves rarely, if ever, rust. But they do become slightly mottled or discolored. To keep them looking attractive, apply a stove polish (blacking) once or twice a year. This comes in liquid or paste form and is easily applied with a rag.

Check inside the firebox now and then and remove soot that

accumulates with a wire brush. Vacuum the ash from the bottom. When the stove is clean, look for cracks that may have developed.

Make sure the door closes tight, so you can't easily pull out a dollar bill sandwiched behind it. If the door is loose, it's probably because the seals around the edge are damaged or worn. Pull them out and replace them. Asbestos seals are traditional; new ones are made of fiberglass. They cost little. Just be certain you buy one of the right thickness. Take the old seal with you to the stove or hardware dealer.

Getting maximum heat from a stove

Obviously you want to get as much heat out of a coal stove as it is capable of giving. You also want the heat to warm not only the room in which the stove is located but also some of the rooms beyond. There are several ways to achieve these aims, and these are by no means all.

Heat reclaimers. A heat reclaimer is a small boxlike device inserted between stovepipe sections just above a stove, furnace or freestanding fireplace. Inside it has a number of metal tubes through which fresh air is blown by a thermostatically controlled fan. As the hot gases from the stove pass around the outside of the tubes, before continuing on up the flue, they transmit heat to the fresh air in the tubes; and the fan forces this heated air into the room or through a duct to another room up to about twenty feet away.

The whole idea of a heat reclaimer is to capture some of the heat that would otherwise be wasted up the chimney and use it to augment the heat you get directly from the stove or furnace and the stovepipe. Manufacturers claim that if the temperature of the flue gases is five hundred degrees, a heat reclaimer can add about nine thousand BTUs per hour to your heat supply. As we said earlier, Cary and Charles use a heat reclaimer with their basement stove, and it does an excellent job of heating their living room.

Used in conjunction with a coal stove that you keep going twenty-four hours a day, a heat reclaimer can obviously add con-

The heat reclaimer installed in the stovepipe captures some of the heat escaping up the flue and circulates it into the house. It can add considerably to the heat output of a fire in a stove, furnace or boiler, but you must make sure that it does not interfere with the chimney draft. (Magic Heat)

siderably to your comfort. With a stove that you use only a few hours of the day, it contributes much less, of course; but it still augments your heat supply without augmenting coal consumption. The one thing you can't be sure about before you make an installation is whether the reclaimer will reduce the chimney draft to a point where the fire in the stove burns less efficiently. We have not had any complaints that it does. But you should not install one if your stove doesn't have a good draft to start with. And if you do install one, you must remove the soot from the exterior of the tubes daily (many reclaimers have a built-in scraper to do this).

Old-fashioned ceiling fans are now widely used, not for summer cooling, but to distribute the hot air that rises to the ceiling throughout the room and back down to floor level. They're especially valuable in rooms with high ceilings. (Hunter)

Other ways to distribute the heat. These methods do not add to your heat supply. They're designed simply to improve distribution of the heat given off by the stove and stovepipe.

1. Since heat rises, the temperature at ceiling level is usually higher than it is at floor level. To break up this blanket of heat and improve comfort where you're sitting, you can install a large ceiling fan like that used in old-time ice cream parlors in the center of the ceiling, or one of the smaller newer fans like that shown. Such fans are especially useful in rooms with unusually high ceilings.

This is one of several kinds of small fans now used to even out the temperature of the air at floor and ceiling levels. Like a large ceiling fan, it also helps to some extent to distribute heated air from the room with the stove or fireplace to adjacent rooms. (Rusth Industries)

Installed high in the wall, this fan is used to transfer heat from the room in which the stove is located to the adjoining rooms. (Preway)

2. Obviously, the more central the location of a stove, the better it heats a large part of the house. We have friends who placed their stove on the hearth of the living room fireplace. The living room is in the middle of their two-story house; the chimney runs up through the stairwell wall. The stove plus the heat from the chimney keep the entire house comfortable throughout the winter.

3. As suggested earlier, one of the most effective ways to distribute heat from a stove when the stove room is walled off from the adjacent rooms is to install high in the wall (or walls) a fan that sucks the heated air out of the stove room.

The same thing can be done less effectively by installing a grilled opening high in the wall.

4. To heat the room (or rooms) above the stove, cut grilled openings in the floor. This was the method used to heat entire houses with old-style gravity warm-air furnaces and it works just as well with a stove. Friends with an eight-room ranch house heat the entire place with a single stove in the basement. They cut one hole in the basement ceiling into the kitchen directly over the stove; a second at the end of the house in the bedroom wing. The holes are covered with grilles and both have small fans which run continuously. The system works so well that the kitchen is often overheated.

Note: Fans draw so little current that you can run one for an entire day for only a couple of cents.

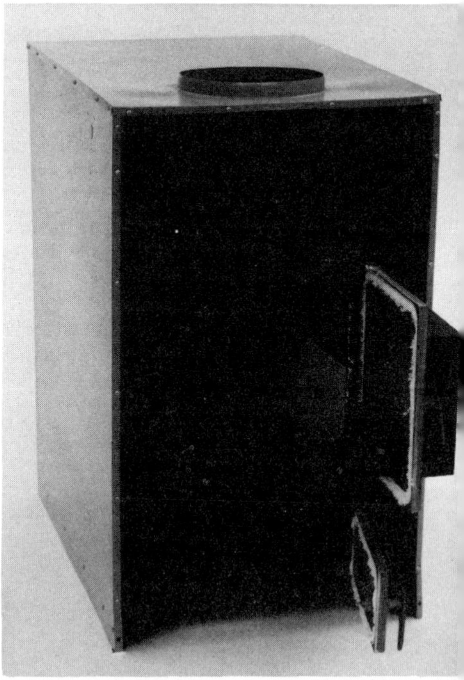

Two small add-on units. At left is a boiler of unusual magazine-feed design in which coal is loaded into a hopper in the boiler from which it feeds by gravity into the fire. (Heathdelle Associates) At right is a furnace that burns either coal or wood. (Suburban)

Chapter 4
Central heating with coal

Only about one half of one percent of American homes are heated by coal-fired central heating plants. But the figure is slowly rising. Every coal furnace and boiler manufacturing firm we have questioned reports that it is hard-pressed to keep up with the demand. Some are back-ordered for a year. One of them, situated in upper New York state, describes the change that is taking place like this:

"When the OPEC trouble started, our sales were concentrated in the eastern coal-mining areas. Now a high percentage of sales is coming from elsewhere–mostly upper New York and New England, where oil was the primary heating fuel. We sell nothing but hand-fired units and 90% of our sales are combination coal/wood heaters. I'll admit that most people buy these to use with wood, but a great many are now switching to coal because it costs no more and requires less work."

Meanwhile, in the Rocky Mountain states–particularly Colorado–coal heating systems are selling about twice as fast as they are elsewhere in the country.

Whether this is a straw in the wind nobody knows. But manufacturers are taking note. Several years ago a Wall Street Journal reporter in Pittsburgh wrote a long account of his efforts to put in coal heating; but he gave the idea up (at that time, anyway) because he said he could find only one coal furnace manufacturer in business, and no stoker makers. Today there are several dozen firms producing coal-heating plants and more than a half dozen stoker makers.

Today's coal furnaces and boilers are basically no different from those made in the thirties and early forties. (Actually, there hasn't been a very startling change in heating plants fired by other fuels, either.) The round firepot has pretty well given way to a rectangular firebox, and along with this change, the external shape of furnaces and boilers has changed from round to rectangular. Certainly the heaters look a lot neater and we're sure they're a lot tighter. They're also more likely to have automatic controls. But they still work in the same way—which isn't meant as a disparagement because they used to work very well. You shoveled in coal a couple of times a day; shook down the ashes and disposed of them and the clinkers via the trash collector. And that was about it. In the late thirties and forties automatic controls began to come in to take care of everything else. Your house was warm, and unless you lived in a soft-coal region, it wasn't unusually dirty. (When Stanley and Elizabeth moved from gas-heated California to an anthracite-heated house in Connecticut before World War II, they were astonished how much cleaner the latter was; but of course, that was in the days before the gas and oil people took real steps to clean up their systems.)

The big difference in coal-fired heating plants today is that there are so many different models to choose from. Whether you're interested in furnaces for warm-air heating or boilers for hot-water and steam heating (but you can forget steam; it's hardly ever used although there are still many old houses that rely on it), you can have either a hand-fired or stoker-fired unit. All the stoker-fired plants are designed specifically for coal, although at least one will also burn oil. But hand-fired units are more diversified. There are those specified for burning coal only (although they can be used for wood); those that burn coal and wood interchangeably; and coal/wood units that can also burn oil or gas or even electricity.

The kind of coal you burn in these plants varies. A few of the hand-fired units are designed specifically for one kind of coal; most can be used with any kind. Stoker-fired units are more likely to be specified for anthracite, bituminous, sub-bituminous or lignite; but many are not.

A problem

Right here you have a problem. The literature put out by most coal-fired furnace and boiler manufacturers is extremely vague and inadequate. This isn't unusual: manufacturers in other fields publish poor literature, too. But the coal people are trying hard to outdo everybody. Consequently, you're very likely to be disappointed with the literature you may request from the producers listed in the back of this book.

You'll have to depend more on the dealers to answer your questions. But this may be disappointing, too. In the first place, finding dealers is difficult because there aren't many of them and the chances are that those you do unearth will be half way across the state from where you live. Even so, if you ask a manufacturer to send literature, be sure to ask also for the names of his nearest dealers.

In the second place, many dealers are pretty short of knowledge about the furnaces and boilers they sell. Posing as potential customers, we've called on several in Connecticut, where interest in coal-fired heating is just kindling. On one particular day we hit one who didn't have the furnaces he sold on display; he had no idea who was qualified to make an installation for us; and until we pressed the point, he didn't know and didn't seem inclined to find out which kind of coal was supposed to be burned in the furnaces. By contrast, the next dealer was right on top of the subject. He showed us his stoker furnace and explained it well. He knew many of the other furnaces and stokers on the market, and when we explained our needs more fully, he said he would immediately dig up a furnace that would suit us better. And he was all set to make the installation even though we lived forty miles away.

Conclusion: If you're really interested in coal heating, don't give up on the first dealers—or manufacturers—you contact. Unlike the industry as a whole, there are some who are definitely eager for your business.

Which coal to use

But getting back to the kind of coal you can burn in a furnace or boiler—

Although many manufacturers indicate that any coal is suitable for their heaters, the fact is that furnace and boiler efficiency is not uniform with all kinds of coal. This is not the fault of the furnace or boiler design but of the quality of the coal used or the firing skill of the operator.

The efficiency of combination furnaces and boilers that burn either coal or wood also varies. If the unit is designed primarily for wood—as most are—it does not burn coal to maximum efficiency. Conversely, if designed primarily for coal, it does not burn wood to maximum efficiency (but of course, this is of little concern if what you want first and foremost is a coal-fired heater).

The same thing is true of stokers. Stokers are designed for either hard coal or soft coal—not both. And there are a few kinds of soft coal that a stoker designed for soft coal can't handle.

In other words, if you're going to put in coal heating, you must first find out which kind of coal is readily available at a decent price in your community. Then you should seek out the heating plant that is designed to burn that coal to best advantage.

How does coal-fired central heating work?

Heating systems are the same whether they burn coal, oil or gas.

Warm-air heating is done with ducts. In a modern forced-air system (almost no one uses the old gravity system), air is heated as it passes through the furnace. From there a large blower circulates it through supply ducts to all parts of the house. It enters the rooms through registers which in cold climates are generally installed just above the baseboards on the outside walls and in warm climates are generally installed high in the walls or ceiling. When the air begins to cool, it is drawn into a return register and carried back to the furnace through return ducts. There it is filtered, heated and circulated through the house again.

Warm-air systems can also be used to humidify the air; and with the addition of an air conditioner, they can be used for cooling and dehumidifying.

In a modern hot-water heating system, water in an enclosed piping system is heated in the boiler and circulated by a small pump to radiators or convectors installed on the outside walls of the house. As it passes through the heating outlets it gives off heat to the room and then returns to the boiler and is reheated and recirculated.

Hot-water systems are not used for cooling, filtering, humidifying, but they can be used to heat your domestic water supply (for bathing, cooking, laundering, etc.) the year round. In fact, most coal-fired boilers either come equipped or can be equipped with a water heater. Known as an indirect or instantaneous or tankless or take-off water heater, this is, in effect, a coil of pipe immersed in the hot boiler water in the top of the boiler. When you turn on a hot-water faucet, the water passing through the pipe is heated by the boiler water. In winter, your hot water is to a great extent a by-product of the house heating system and costs you little. In summer, however, you must continue to keep a fire going in the boiler in order to have hot water for bathing, etc. This doesn't cost as much as you might expect, because the fire is kept low enough to heat the water without heating the house. But it means that you don't get year-round hot water for free (as some people seem to think).

(If a coal-fired boiler is not designed for a water heater or if you have a coal-fired furnace, the best way to heat your domestic water supply with coal is to buy a separate water heater with a you have a coal-fired furnace, the best way to heat your domestic water supply with coal is to buy a separate water heater with a storage tank. These are often called pot boilers because of their small size and shape. But unfortunately, we have not been able to find a source for them in this country. You'll have to go to Europe for them.)

Coal-heating basics

1. The size of the coal burned in a central heating plant influences the efficiency of the unit. Anthracite, however, is the only coal with a uniform sizing scale. Before leaving the mine, it is graded into one of the following categories:

Broken	over 3¼" diameter
Egg	" 2 7/16" "
Stove	" 1 5/8" "
Nut	" 13/16" "
Pea	" 9/16" "
Buckwheat	" 5/16" "
Rice	" 3/16" "
Barley	" 3/32" "

Of these, the sizes generally used for small furnaces, boilers and stoves are stove, nut, pea, buckwheat and rice. The first three are for hand-fired equipment. Buckwheat and rice are the commonly used sizes in stoker-fired equipment.

Soft coals are not graded uniformly because it is not so important to burning efficiency to have all pieces of similar size. As a matter of fact, all sizes come mixed with some "fines"—dust-like particles—to improve burning. Nevertheless, stokers require smaller coal than hand-fired heaters.

In using a furnace or boiler for the first time, it's best to start out with the coal size specified by the manufacturer. But later you may find it advisable to switch to another size. For example, if the draft is inadequate, you may be able to offset it by using a larger coal. Conversely, if a large coal doesn't burn completely, you may correct the problem by changing to a smaller coal.

(Obviously, if you want to test the effect of different sizes of coal, you should buy it by the bagful, not by the ton.)

2. Unlike oil and gas-fired heating systems, in which the fire goes on and off as the thermostat demands, the fire in a coal-fired

system burns constantly and heat output is controlled largely by introducing more or less air.

3. According to the United States Environmental Protection Agency, the average stoker-fired anthracite furnace or boiler burns a total of 6.7 short tons per year. (A short ton equals two thousand pounds.)

The average stoker-fired bituminous heater burns almost six tons.

The average stoker-fired lignite heater burns 10.9 tons.

Coal consumption of hand-fired furnaces and boilers cannot be measured so easily because it is variable. But it generally runs a trifle lower.

Hand-fired furnaces and boilers

Hand-firing a furnace or boiler usually doesn't mean that you have to check it frequently during the day. In fact, when coal was the standard heating fuel, almost all home and small business owners soon reached the point where they could run their heating systems pretty much by remote control. They did this either by learning exactly how the fire should be fed and how much the drafts should be opened, or they rigged up a system of chains to control the drafts from upstairs. Today, there are a few very simple furnaces and boilers that have to be controlled in this way, but most come with automatic controls. Thus all you have to do is feed the fire a couple of times a day, bank it at night, and shake down and remove the ashes and clinkers.

There are three kinds of hand-fired heating plants: (1) Coal only; (2) coal/wood, or dual-fuel units; and (3) multi-fuel units that burn either coal or wood and also oil, gas or electricity. There are also wood-only heating plants that are practically identical to the coal-only plants. We mention these only to make the point that a wood-only unit should never be used to burn coal because it isn't built to withstand coal's higher heat output. (But you can burn wood in a coal-only heater or a dual or multi-fuel heater.)

Not too many coal-only units are on the market. A few are designed for use in old-fashioned gravity warm-air heating systems; the others are for modern forced-air or forced-hot-water systems with a circulating pump or blower.

Coal/wood units are very common and essentially similar to the coal-only units. The main difference is that they are designed to give best results when fired with wood.

Both types of plant are designed to be free-standing or to be tied into an existing oil or gas-fired heating system. The latter are commonly known as add-on units but this doesn't mean they can't be used by themselves or that the free-standing units can't be used as add-ons. The only difference is that the add-on units are externally designed for placing cheek-by-jowl with an existing heating system whereas the free-standing units generally are separated from an existing heating system.

This brings up a point that merits some emphasis: To enjoy the economy of coal heat, you don't have to rip out and scrap your present furnace or boiler. Many, many people who are buying coal heaters are using them as auxiliary or supplemental heating plants. To cut fuel costs, they use their new coal-fired units most of the time (especially during the coldest months) and they switch over to their old oil or gas-fired plants only when they are going to be away from home for any length of time and can't stoke a coal fire, and during the fall and spring when they don't need heat twenty-four hours a day. Some systems make this changeover automatically; thus if you accidentally let the coal fire burn down, the old plant automatically comes on to keep your house warm.

A further advantage of a two-fuel system is that, if you have a power outage, the coal-fired unit will help to keep your house above the freezing mark for some time. However, since a blower that circulates warm air and a pump that circulates hot water run on electricity, circulation of heat is sharply reduced (the heat moves through the ducts or pipes by gravity only). Building up the fire to compensate partially for this is verboten because it can result in damage to the heater.

In this multi-fuel boiler, coal or wood is burned at this end; an oil burner with a separate combustion chamber is at the opposite end. A gas burner can be substituted for the oil burner. (Oneida)

Teaming a coal furnace or boiler up with an oil or gas furnace or boiler does not entail great alteration of the entire heating system. It simply means that the ductwork or piping immediately above the heating plant is changed so that both furnaces or boilers are hooked up to it. Thus the heat entering your rooms comes from whichever unit happens to be on at the time. This interconnection of the two heating plants can be made by any competent heating contractor but it is a complicated job.

Multi-fuel units give you the same flexibility—the economy of heating with coal during the winter and the ease of heating with oil or gas during fall and spring and when you go away. But multi-fuel furnaces and boilers are incorporated in a single jacket. In most cases, these units have two separate combustion chambers—one for

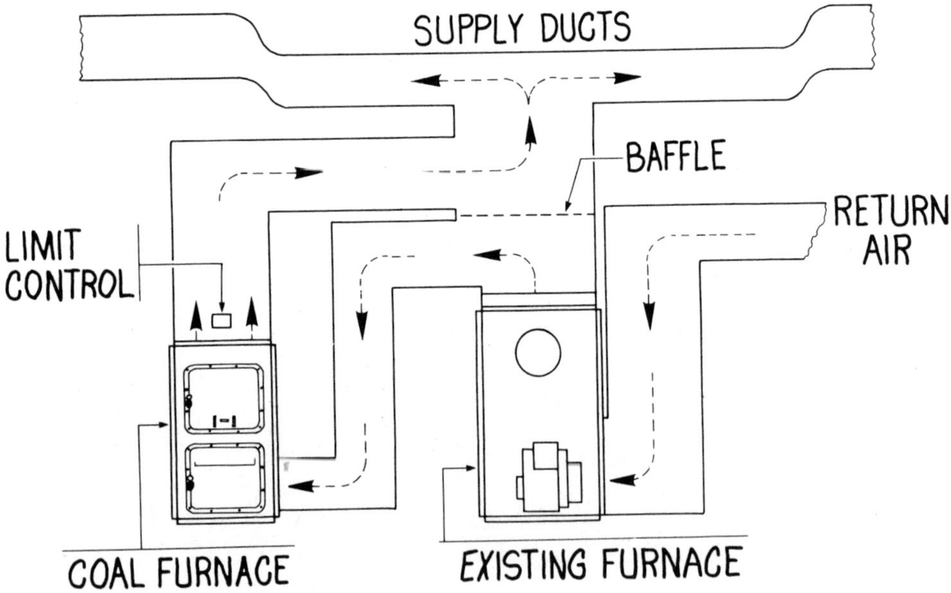

If you have a perfectly good oil or gas-fired furnace, you can have a coal-fired furnace teamed up with it so you can heat the house with either one. Coal-fired boilers can be connected with existing oil or gas-fired boilers in the same way.

coal and wood; the other for oil or gas. In a few units, however, all fuels are burned in the same combustion chamber. This reduces the over-all size of the heater but also results in somewhat lower efficiency. In either case, the heater is designed so it switches over automatically from coal to oil or gas if the coal fire fails to maintain room temperature. It can also be switched from one fuel to the other as you wish. And it gives you some heat during power outages.

Obviously, multi-fuel units are installed only in new houses or stores or when you decide to junk an old heating plant. They cost more than twice as much as a coal-only or coal/wood heating plant—roughly twenty-three hundred dollars versus nine hundred dollars for heaters of comparable size. (These are furnace prices. Boiler prices run quite a bit higher because of differences in materials and design.)

Running a hand-fired boiler

It takes time to learn to run a hand-fired coal boiler or furnace. For one thing, the operation is different from anything you've known before. For another thing, each boiler (or furnace) and each installation has its own peculiarities. This is why we can't be terribly specific. You must go through the whole business with your installer and/or dealer; compare what they say with what the manufacturer says in his operating manual; and then reconcile the differences with the dealer.

Two kinds of boiler are illustrated. The first is a more or less standard unit much like those built in the past. It operates on coal and can be used for wood if you must.

In the drawing, you see the boiler from the side. The firebox is loaded through the upper door on the left. The grate is shaken by a long vertical handle to the left of the doors; and the ashes are deposited in the ashpit, from which they are removed via the bottom door on the left. The ashpit damper through which the primary draft air enters the boiler, under the grate, is at the bottom on the right. It is connected by a chain to the barometric draft regulator on the flue and the damper control arm. The mushroom-

A hand-fired coal-only boiler. The way it's controlled is shown in the drawing. (Slant/Fin)

shaped unit below the control arm is the aquastat, which moves the damper in response to changes in water temperature.

When the aquastat senses that the temperature of the boiler water drops below a predetermined point, the control arm automatically opens the draft and simultaneously throttles down the barometric draft regulator. This forces the fire to burn more briskly, raising the water temperature until the aquastat decides it's high enough and actuates the control arm, which then closes the draft and opens the draft regulator.

Throughout the entire operation, while the fire is burning fast and also while it's burning slow, secondary air is admitted into the combustion chamber above the fire through the adjustable vent in the loading door. This secondary air is necessary to burn the gases given off by the fire, thus giving you the maximum heating value

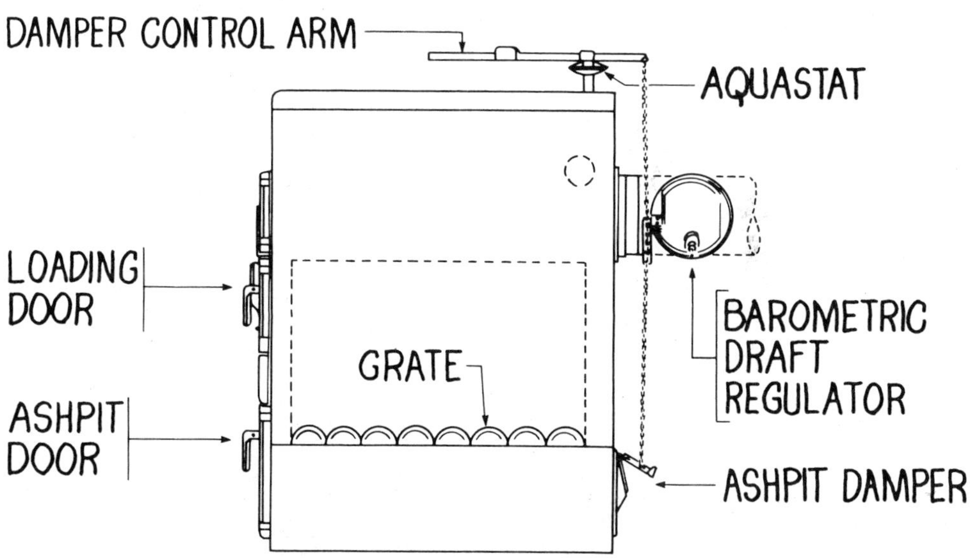

This is the first hand-fired boiler described in the text. The aquastat senses the water temperature of the boiler and opens and closes the drafts accordingly.

out of the coal and reducing the emissions up the flue to the outside air.

The circulation of hot water from the boiler to the radiators is controlled by a room thermostat located in the living area of the house. This is not shown in the drawing because it does not control the boiler dampers. Instead, it controls the circulating pump installed in the piping system outside the boiler. In this respect, it is no different from the thermostat used with an oil or gas-fired boiler. When the house temperature drops, the thermostat tells the circulator to turn on; then it later tells the circulator to turn off.

The second drawing is a side view of another type of boiler which, in this case, burns coal or wood and, if you wish, oil or gas.

When the aquastat (11) senses that the boiler water temperature is too low, it activates a blower (1) that forces primary combustion air into the combustion chamber. When you're burning coal and the fuel-selector damper (4) is pushed in, the air is directed up through the grate and the fire. (When the fuel-selector damper is pulled out for burning wood, the primary air enters the combustion chamber through the air ports at 5.)

The direct-draft damper (12) in the top of the combustion chamber is opened only when the fuel door is opened to permit smoke to bypass the combustion flue and prevent it from entering the basement.

Secondary air enters the boiler at (2), is preheated and passes through the secondary air distributor (7) into the combustion flue (9) where more complete combustion of the hot flue gases occurs. The hot gases then travel through the heat chamber (8), transferring additional heat to the water (10).

The barometric draft regulator (6) controls the chimney draft.

The loading door (13) is to the right of the oil or gas burner (14). A switchover aquastat automatically turns the burner on if the coal fire does not maintain the proper water temperature.

As in the case of the first boiler, the circulating pump that delivers hot water to the radiators is controlled by a room thermostat. Neither the thermostat nor circulator is shown.

Whenever you start a fire, leave a layer of ash on the grate to protect it from the high heat and to keep the coal from falling through. (When starting the fire for the first time in a clean boiler, substitute about three inches of coal for the ashes.) Open the drafts. Place paper and kindling on the grate and when the wood is burning well, toss in a thin layer of coal. Never, never, never use kerosene, gasoline or other fluid to start a fire either in a central heating plant, stove or fireplace.

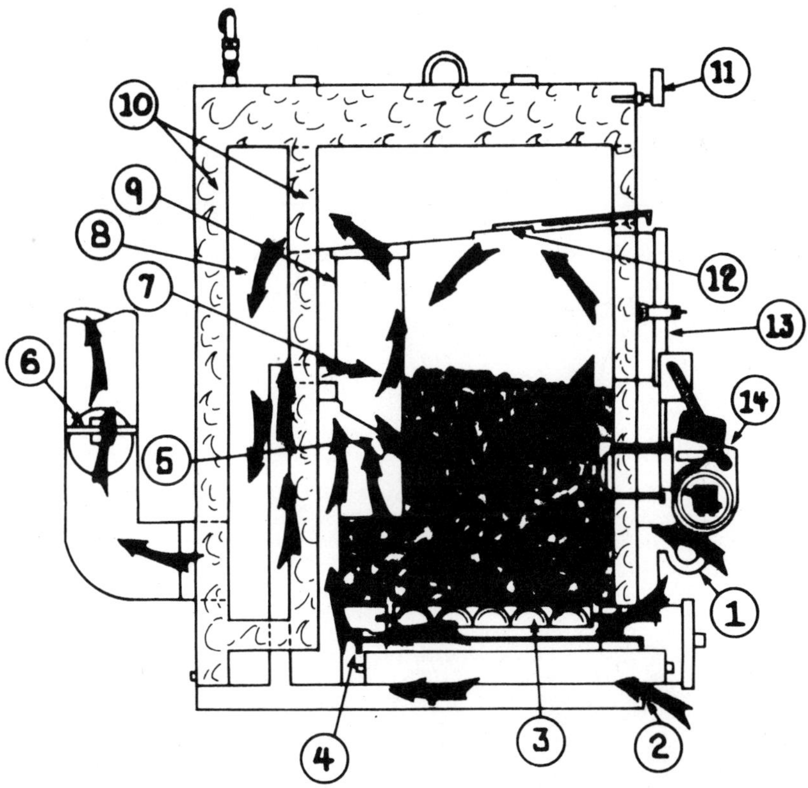

This is the second hand-fired boiler described. It, too, is controlled by an aquastat. It differs mainly from the first boiler in that it has a blower to introduce primary air. The direction of air flow is changed if wood instead of coal is burned. The arrows show the air flow for both a coal fire and a wood fire.

As soon as the coals are burning strongly, add more. Then more until you have a deep fire bed level with the bottom of the loading door. Leave some red coals visible after loading. This is particularly important when using smaller sizes of coal. The red coals are usually pulled to the front of the firebox and the fresh coal is added in back (some manufacturers recommend putting the red coals in the back of the firebox), but the same effect can be achieved either by piling the fresh coal around the edges of a central bright spot or by heaping fresh coal on one side and alternating sides with each successive firing. The latter method, called side banking, is particularly recommended for bituminous coal.

Sub-bituminous and lignite coal can be fired in the same ways but the central-cone method is generally preferred. In this case you pile the fresh coal up in a cone in the center of the fuel bed. The top of the cone is level with the middle of the loading door.

About twenty minutes to an hour after a boiler is fired with one of the soft coals, run a poker into it gently to break it up. But be careful not to bring the ash up to the top of the fire, where it melts and forms clinkers.

Daily operation of the fire follows a pattern such as this:

In the morning, get the fire going briskly by admitting air up through the grates; then heap on coal.

During the day, the automatic controls should take care of things but until you feel you're adept at running the fire, it's a good idea to check it several times.

At night, remove from the ashpit the ashes that were shaken down the previous day. If they're hot, you can dampen them. In any event, dump them into a covered steel ash can. Then shake down the ashes produced during the day. Don't be over-enthusiastic about this. Use a short, choppy motion on the lever. Stop shaking as soon as one or two pieces of live coal fall into the ashpit or as soon as a red glow appears on the grate. Lift out clinkers with tongs and put them in the ash can. Then load in coal to the level of the fire door and disengage the automatic draft controls (i.e., unhook the chain on the damper control arm of a standard heater) so there is just enough air entering the firebox through tiny cracks around the various doors to maintain combustion. This is called banking. Contrary to popular opinion, banking does not mean sloping the coal up toward the back of the firebox; instead, it means conserving heat, which is done by reducing draft.

If a fire burns down to the point where it threatens to go out, don't poke or shake it. Open the drafts and spread a thin layer of fresh coal over the top of the entire fire. When this is well lighted, shake the grate and add more fuel.

The worst mistake you can make in operating a coal fire of any kind—but especially a coal fire in a boiler—is to over-fire it by loading in too much coal and producing too hot a fire. A boiler may explode in such circumstances or the walls of the combustion chamber may be cracked.

The first way to protect against this calamity is to install a properly sized pressure-relief valve that vents hot water out of the boiler when it is overheated. This valve is an absolute necessity.

The second preventive measure is to have a high-capacity unit heater installed in conjunction with the boiler. A unit heater is a device you've probably seen in the service bays of gasoline stations and in commercial garages. Mounted up near the ceiling, it's a rather large, thin, rectangular frame with a fan. Hot water from a boiler located elsewhere in the building circulates through it in pipes. When used for heating purposes, the fan blows air across the pipes into the garage. But when used to protect a boiler against emergencies, the heater is aimed through a wall to the outdoors. The heater piping is connected to the boiler in such a way that, if the temperature of the boiler water begins to rise dangerously, the heater turns on automatically and rapidly blows the excess heat outdoors.

There are alternatives to the unit heater, but the simplest is, unfortunately, the least practical: Just stick around the house so that, if you suddenly notice the house temperature shooting up, you can run down to the basement, shake the fire into the ashpit and shovel out the hot coals as quickly as possible.

Power failures pose a problem because the circulating pump

stops working: and if the fire is burning briskly, there is no way to carry off the heat (even if you have a unit heater, because it can't run either). So the boiler temperature begins to rise sharply.

Happily, the aquastat will close the draft when the boiler water temperature reaches a preset point; but if you've overloaded the firebox with coal, the temperature will continue to rise for awhile—to perhaps a dangerous level. This will cause the pressure relief valve to start venting, or "blowing." But you still may be in trouble. What to do?

You should immediately race to the basement and close the drafts yourself. Then dump the fire into the ashpit.

During a power outage, a coal fire must be kept low. You don't want to extinguish it because in winter, even though the circulator is not running, some of the heat gets up into the house and helps to keep it warm. As pointed out earlier, that's one of the advantages of coal-fired heating systems. But keep the fire low and open the check valves and thermostatically controlled valves in the piping system to let hot water flow through the pipes and radiators as best it can.

Hand-fired coal furnaces

Hand-fired coal furnaces are available in the same types as hand-fired boilers—coal-only, coal/wood and multi-fuel. One manufacturer also sells coal/oil furnaces. They're made for free-standing and add-on installation. Of course they differ from boilers in that the combustion-chamber heat is transferred to air rather than water; therefore, they're constructed in quite a different way and cost less. But they operate in pretty much the same way.

A furnace is controlled by a room thermostat installed in the living area of the house. In most cases, when the thermostat calls for heat, it signals a small damper motor that is connected to the ashpit door and barometric draft regulator by an arrangement of chains. The motor opens the ashpit door and closes the barometric draft regulator, thus increasing the draft through the fuel bed and causing the fire to burn more vigorously. At the same time, the thermostat turns on the blower that circulates heated air through the ducts to the registers. Then, when the house temperature has risen, the thermostat tells the drafts to close and the blower to stop.

A hand-fired coal/wood furnace for a forced warm-air heating system. When the house thermostat signals, the damper motor opens or closes the drafts.

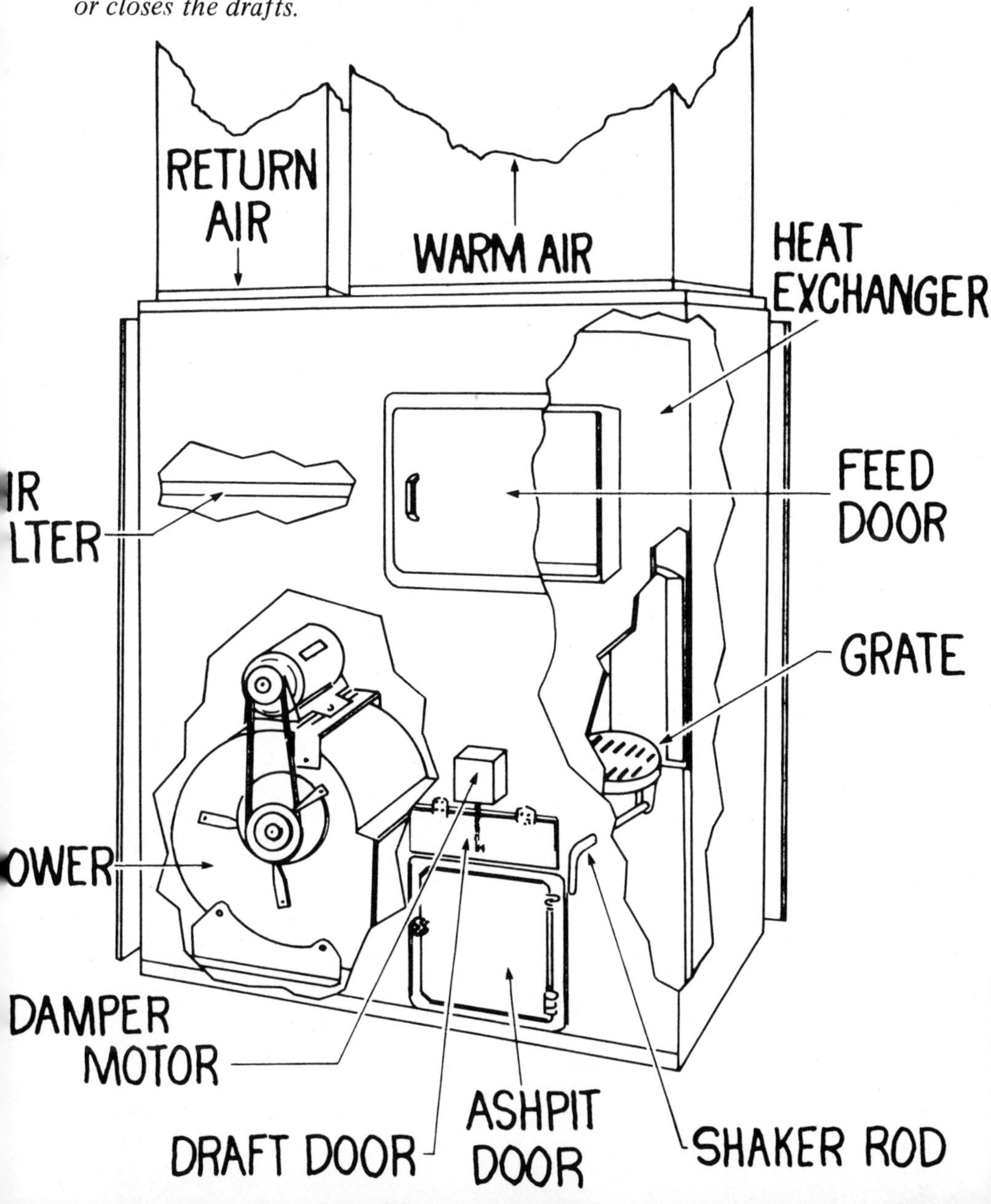

To prevent damage, the heating plant is also equipped with a limit control (high-limit switch) and a blower control. The former checks the air temperature in the furnace plenum, and if this rises dangerously, the switch overrides the thermostat and closes the furnace drafts until the air temperature returns to normal. The blower control works in much the same way when the thermostat is not calling for heat but the furnace temperature for some reason builds up too high. If this happens, the control starts the blower, which forces the excess heat into the house.

For how to fire and manage a furnace, follow the directions under boilers.

Stoker-fired furnaces and boilers

Stoker firing gives you almost the same automatic control of your heating system that you get with gas or oil. You pay more for an installation (a 100,000-BTU stoker-fired boiler costs roughly thirty-three hundred dollars before installation), but day-to-day operating costs are less than with oil but a shade more than with a hand-fired coal system. Gas is still the cheapest fuel.

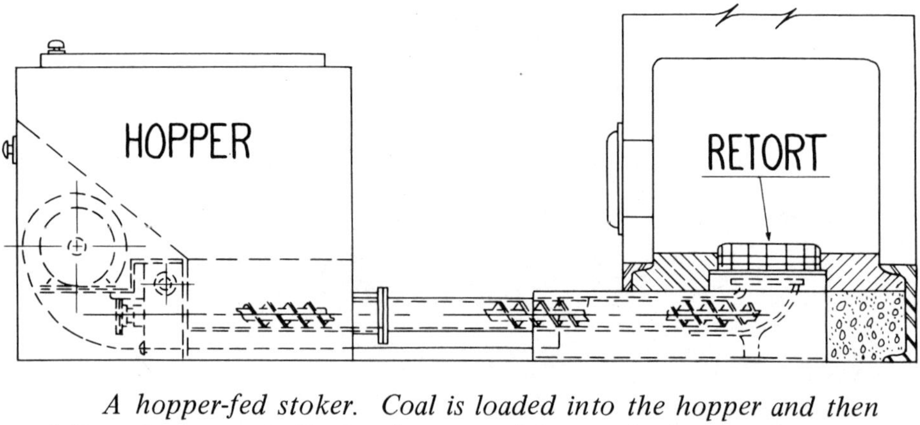

A hopper-fed stoker. Coal is loaded into the hopper and then delivered automatically by the worm-drive mechanism to the retort in the combustion chamber (in this case, an under-feed unit). The stoker also controls the air supply.

The actual amount of work involved in running a stoker-fired furnace or boiler depends on the system. If the stoker is designed to feed coal directly from the bin to the heater, the only thing you have to do is shovel the ashes out of the ashpit and remove the clinkers. If the stoker is hopper-fed, you must occasionally shovel coal from the bin into the hopper in addition to removing ashes and clinkers.

Use of a stoker produces more uniform comfort conditions in the house because the rate of firing is more even. And a more uniform firing rate probably results in fewer emissions of undesirable gases and particulate, although we have not run across any studies to prove it.

Stokers are designed to burn either anthracite or bituminous or sub-bituminous and lignite coals. The coals should not be used interchangeably. Broadly speaking, most of the current stoker-fired

This boiler is equipped with a flat, stationary-rate stoker. The stoker pushes coal onto the sloping grate, where it burns. As fresh coal is added, the ashes drop off into an ash can in the ashpit. The small coils in the top of the boiler are the domestic water heater.

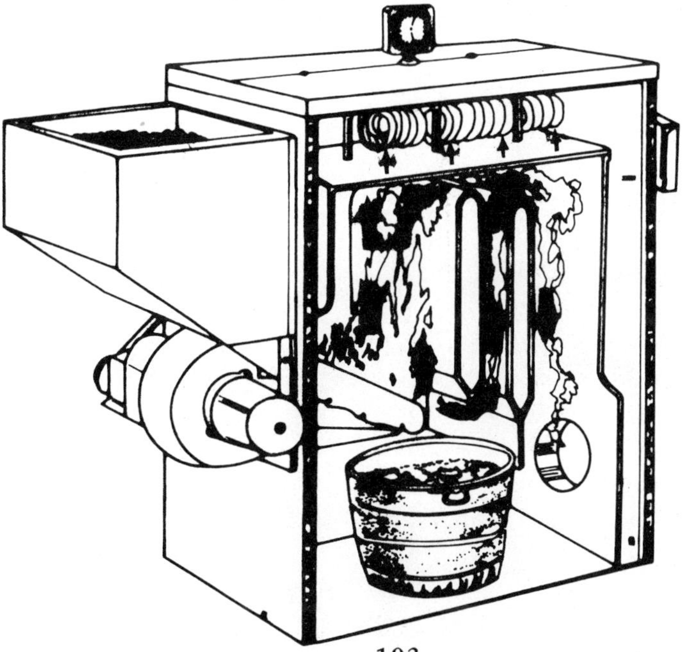

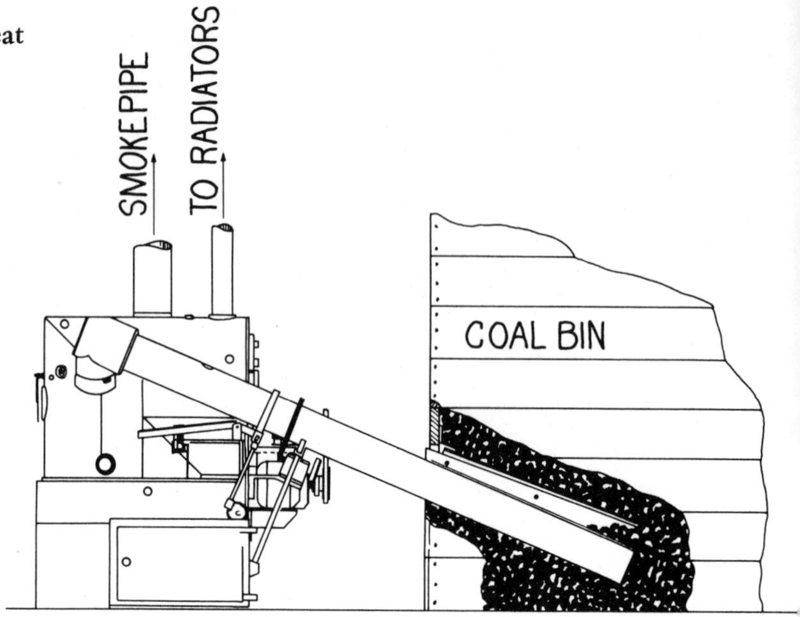

A bin-feed stoker delivers coal directly from the bin to the heating plant, which in this case is a boiler of the uncommon over-feed type.

In the boiler shown above, the coal drops into the retort. The weight of the firebed forces the ash onto the grate. As the grate moves slowly forward, steps on it cut ash from the bottom of the firebed and push it toward the ash can. Then as the grate moves back, the ashes fall into the can.

boilers are designed for anthracite and the stoker-fired furnaces for one of the soft coals; but there are furnaces that use anthracite and boilers that use soft coal.

The stokers are installed partially inside and partially outside the heating plant. Powered by electricity, they automatically feed small sizes of coal into the combustion chamber of the furnace or boiler, where it is ignited by the on-going fire, consumed and then automatically dumped into the ashpit. The majority of stoker-fired heaters are so-called underfeed units. In these, the coal comes up into and is burned in a bowl-shaped retort that is set into a wind box that delivers air to the fuel bed through holes called tuyeres. In overfeed units, which are rare, the coal is dropped into the retort. In both of these units, the coal is conveyed from the storage compartment to the combustion chamber by some variety of worm drive.

A third kind of stoker is a flat, stationary-rate unit with a hopper mounted high on the side of the furnace or boiler. A motorized feed mechanism pushes the coal downhill onto a sloping grate, where it is burned. As fresh coal is added, the ashes of the preceding load are pushed off the end of the grate into the ashpit. This type of system is much like the old magazine-fed furnaces in which coal was gravity-fed into the combustion chamber.

Whatever the design of the stoker, operation of the heating system goes about like this:

When the room thermostat calls for heat, the stoker feeds coal and air into the combustion chamber. At the same time, the circulator that forces heated air or water into the house is activated. Then, when the house temperature has risen, the entire system shuts down.

The rate at which coal and air are fed into the combustion chamber is set at the time the system is installed but can be adjusted any time thereafter. During the average portion of the heating season, a stoker operates for twenty minutes and is off for forty minutes; but in extremely cold weather it may run continuously for twenty-five minutes, then turn off for five minutes to allow the fuel bed to cool and to prevent forming of large clinkers. In warm weather, on the other hand, the stoker may run for only five minutes out of every thirty, or just enough to maintain the temperature of the domestic water supply circulating through the water heater coil in a boiler.

In addition to the thermostat, a stoker-fired heating plant incorporates a limit control, hold-fire control and barometric draft regulator.

The limit control stops the stoker if the boiler or furnace pressure gets too high.

The hold-fire control is designed to keep the fire burning in the firebox in mild weather, when the thermostat is not calling for heat. This control may be either a timer that you adjust to any desirable on/off frequency or a stack thermostat, which automatically activates the stoker when the fire dies too low. By and large, one is as good as the other. But if you live in an area with frequent power outages, the stack thermostat is recommended because, when power is restored, it prevents the stoker from feeding fresh coal into a fire that has gone out.

The barometric draft regulator helps to control the draft through the fuel bed.

To start a fresh fire in a stoker-fed heating plant, open the chimney damper; set the room thermostat higher than the house temperature; open the draft that supplies secondary air to the firebox; run the stoker until it partially fills the retort with coal; make a hollow in the middle of the coal and build a wood and paper fire in this. When the kindling is burning freely, scrape coal in around (not over) it. Then as the coal starts burning well, adjust the manual drafts as necessary and let the automatic air control on the stoker take over. Some adjustment of the air supply and coal feed may be necessary, in which case, follow the manufacturer's directions; but since these feeds are set at the factory or by the dealer, you usually don't have to change them unless you change fuel sizes or have fiddled with them before.

The ash deposited in the ashpit should be removed regularly because it can upset combustion and damage the heater if it becomes too deep. In some cases, the ash is dumped into a removable pan or can that eliminates need for shoveling. At least one stoker firm sells an automatic ash-removal device.

Ashes produced by most stoker-fired boilers and furnaces are removed by hand. This is an automatic ash-removal unit available as an optional accessory from Prill Manufacturing Company.

Clinkers must be removed as they form by turning off the stoker and letting the fire cool for several minutes; then carefully raising the clinkers to the top of the fire with a poker and lifting them out with tongs. In cold weather, clinkers may have to be removed once a day.

The stoker itself should be lubricated several times during the year and if it should grind to a halt because of some obstruction in the drive mechanism, this must be removed. (The stoker is equipped with overload protection devices that shut off the power if something goes wrong with it. After correcting whatever caused the shutdown. you can restart the stoker by pressing a reset button.)

The frequency with which you must shovel coal into the stoker hopper depends, obviously, on the size of the hopper. Some hold as much as six hundred pounds. If you prefer a bin-feed stoker, of course, no shoveling is necessary. However, as the bin empties, you have to rake the coal forward so that the feed screw can pick it up.

When loading a hopper, try to mix very small coal with the larger coal in the proportion of 15% to 85% so that the fire will burn at top efficiency. (When a coal bin is filled, the large coal tends to run to the sides of the pile. So by taking one shovelful from the middle of the bin and one from the sides you get about the right mixture.)

Installing a coal-fired heating plant

There's one fact of life you must recognize before you get too interested in a coal-fired heating plant, whether hand-fired or stoker-fired: It may not be adapted to your home or business unless you make alterations—perhaps very extensive alterations—in the building.

In the first place, you have to have a coal bin that's large enough to hold about a half year's supply of coal. This is because it costs less to buy coal in large quantities and it helps to guard against your running completely out of coal in case of a strike or some other emergency situation.

Since a ton of coal takes up a space measuring approximately 2 x 4 x 5 feet, you need a 4 x 5-foot bin six feet high to store three tons. In addition, you need space in front of the bin to wield a shovel; and you need additional space measuring roughly thirty-five square feet for the heating plant itself.

In the second place, your house or business premises should be so placed and laid out that the coal dealer can back his truck up to it and chute the coal directly into the bin. This is for reasons of economy and also because we're not at all sure how many coal dealers employ men who are willing to fill bins sack by sack or wheelbarrow-load by wheelbarrow-load.

In the third place, all coal-fired furnaces are of the so-called upflow design, meaning that the heated air is discharged from the top of the furnace. This means that, unlike gas-fired downflow furnaces, the duct system cannot be installed in a concrete slab or crawl space below the furnace: it is above the furnace.

(Coal-fired boilers, like boilers using other fuels, can pump water to radiators overhead as well as through a slab or crawl space to radiators on the same floor level as the boilers.)

From these three requirements, you can get a good idea of the kind of house or business place you should have for coal heating.

1. It should be close to the street or driveway.

2. The bin should be on the street or driveway side. And, of course, the furnace or boiler should be on the same side since you don't want to carry coal by the shovelful for any great distance and there's a limit to the length of stoker feed screws.

3. The heating plant should be in the basement, where there is ample room for the coal bin and where the bin is below the coal-chute outlet of a coal truck. This location is also necessary if you install a furnace; otherwise the heat can't be delivered to the registers in the house.

4. If you install a boiler, the house can be built on a slab or over a crawl space. But there still must be adequate room for the bin and it should be below the delivery truck's chute.

If your house is adapted to coal heating, your next step is to find a dealer who really knows the business of installing coal-fired systems. You need him to determine what size plant you require; to advise you on the plant that best meets your needs; and to make the installation <u>in compliance with all pertinent building codes</u>. These are things the average home owner or small business owner cannot do himself although he ought to have some idea about them so he can check up on the professional installer who may not know as much as he pretends to.

Chimney. The flue size required is specified by the furnace or boiler manufacturer.

The chimney into which the heating plant is vented can be built of masonry with an approved flue lining or of insulated Class A "all-fuel" metal pipe. If your chimney is metal, it must not be placed closer than six inches to any combustible material on the outside of the house. If it runs up through the house, it must be enclosed by noncombustible materials spaced at least twelve inches from it. If it extends into the room in which the heating plant is situated, there must be an eighteen inch clearance between it and all combustible materials.

Furnace or boiler clearances. A heating plant should be located to provide a short, direct connection to the chimney and with sufficient room for you to get at all sides easily.

The table shows the clearances required from combustible surfaces unless the manufacturer specifies otherwise.

	Hand-fired furnaces	Stoker-fired furnaces	Boilers
Front of heater	48"	48"	48"
Sides and back of heater	18"	6"	6"
Top of heater			3"
Top and sides of plenum	18"	6"	
Pipes			1"
First three feet of supply ducts	18"	6"	
Second three feet of supply ducts	6"	6"	
Supply ducts six feet beyond plenum	1"	1"	

If you add a coal-fired furnace to an existing oil, gas or electric furnace, allow eighteen inch clearance between a combustible surface and the add-on and the duct connecting into the plenum of the existing unit.

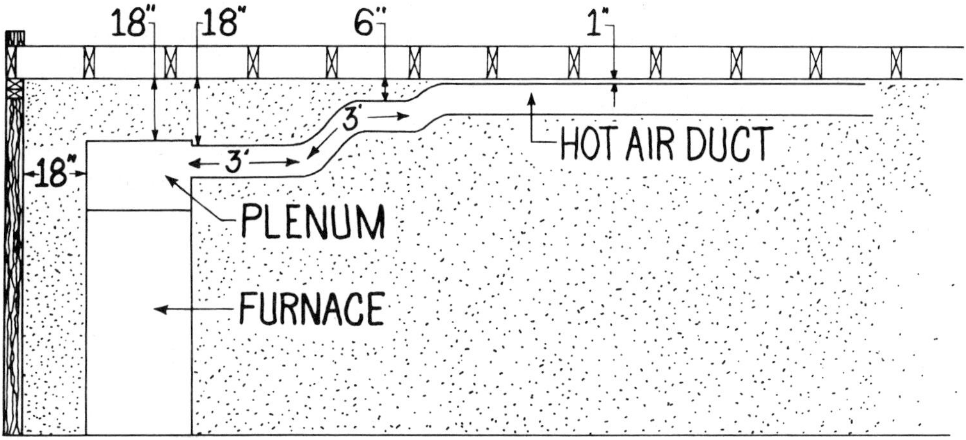

Adequate clearances must be allowed between a furnace or boiler and all combustible surfaces. The clearances shown are for a hand-fired furnace. Stoker-fired furnaces and boilers require less clearance.

These clearances can be reduced somewhat if the combustible surface is covered with noncombustible material as specified in the preceding chapter.

Since most furnaces and boilers are installed on a concrete slab, no special floor protection is required, although if your basement often floods, it's advisable to raise the heating plant two to three inches. If the floor is combustible, however, it should be covered with four inch thick hollow masonry blocks laid so that air can pass through the cores from one side of the protector to the other; and the blocks should be covered with twenty-four gauge sheet metal. The protector extends eighteen inches in front of the heater and to either side of the loading door, but can be flush with the sides and back of the heater.

Chimney connector clearance. If the furnace or boiler is connected to the chimney by a single-wall twenty-four gauge steel pipe (the usual arrangement), the minimum clearance between it and a combustible surface is eighteen inches. The connector should be sloped slightly upwards toward the chimney.

Coal Heat

Safety measures. The devices used with boilers and furnaces to prevent serious problems have been described. Just don't overlook them.

Combustion air. In a very tightly constructed house, it often happens that there is not enough incoming fresh air to maintain good combustion in a coal fire. If this seems to be the case in your house—or if it proves to be the case after you've been running the heating plant awhile—a fresh air inlet of four inch or greater diameter should be provided in the wall of the furnace room. This, of course, may cool off the room more than you like, in which case, install a duct from the inlet to a point close to the heating plant.

Maintenance

The heat-distribution parts of a coal-fired heating system require the same maintenance as they do in any other system. That is, you should make sure that the circulating blower or pump is lubricated; that duct joints are tight; that the filters are cleaned or replaced frequently; and so on.

Maintenace of the coal-fired furnace or boiler itself is as follows:

1. During the heating season, check inside the heater every month to make sure there is not too much build-up of soot, which keeps the heating surfaces from absorbing the heat given off by the fire. When cleaning is necessary, use a wire brush to brush the soot down into the ashpit. A vacuum cleaner is a big help.

2. Take a look now and then at the chimney connector (smoke pipe). If the joints are loose, tighten them. If there are holes in the pipe, replace the pipe at once.

3. In the spring, whether you shut down the plant or use it to supply domestic hot water throughout the summer, clean the furnace or boiler thoroughly. Remove the chimney connector and wire-brush it inside. Unless you do this, the summer humidity may mix with the soot and fly ash and speed corrosion of the plant. Since chimney connectors seem to be particularly susceptible to

corrosion, it may be wise to wrap them in plastic and store in a dry place till fall.

Check the chimney for soot build-up, too.

Thorough-clean the grate and ashpit.

Completely fill a boiler with water to prevent internal corrosion.

4. Have a look at the firebrick lining of the combustion chamber during the thorough spring cleaning. It may need attention.

5. Lubricate the stoker according to the manufacturer's directions. In the spring, clean and repaint the hopper; clean siftings out of the wind box; and clean the tuyeres.

6. Be sure at all times that the loading, ashpit and draft doors fit tight. If when you hold a candle flame next to the edges of a door the flame is bent, there's an air leak and the door gaskets should be replaced.

You should also check the tightness of the clean-out door at the base of the chimney at least once a year, because if it's loose, cold air will be drawn into the chimney and reduce draft.

Since a poorly maintained heating plant is a threat to your comfort and pocketbook, we recommend that you employ a professional to inspect and clean it in the spring. Just turning the dirty job of cleaning the heater over to somebody else is reason enough for this modest expenditure.

Heat reclaimers

The same sort of chimney heat reclaimer that is used with stoves can be used with a coal-fired furnace or boiler. Because the fire is burning constantly, there is usually enough heat in the flue for the reclaimer to extract and circulate into the basement or via a duct, to a nearby room; and when the fire is brisk, the heat savings can be substantial.

But there is danger that, if the chimney flue is only marginally large enough for the heating plant, a heat reclaimer will reduce the

This ancient coal-fired boiler was converted to oil but can easily be reconverted to coal. The cost of converting modern oil or gas-fired boilers or furnaces to coal runs very high, however.

draft so the plant doesn't operate efficiently. (This can also happen if you use a heat reclaimer with a stove; but since a stove is located in the living area, you're much more likely to notice that it is not going well.) For this reason, you should not install a reclaimer unless the dealer or other competent heating man examines your heating plant and gives you an OK.

Converting existing heating plants to coal

A modern gas, oil or electric-fired furnace or boiler cannot be converted to coal without such extensive reconstruction that the project is of doubtful value. A conversion should be considered only if you have an old coal-fired plant that has been converted to oil or gas. In this case, it's an easy, inexpensive job to switch back to coal (provided you can find the necessary grates), and if you want to put in a stoker, go ahead. (Almost all hand-fired coal heating plants—whether they were once converted to another fuel, or are still being used for coal—can be equipped with a stoker.)

Even then, installing a new coal or coal/wood heater in conjunction with your existing plant probably makes more sense because it will be more efficient than a conversion unit and will give you the flexibility of a multi-fuel system.

Space heaters are small enough to be installed in any room but have sufficient BTU output to heat an entire small house. Stoker-fired heaters like this are common. A blower circulates the heat produced. Hand-fired space heaters heat largely by radiation.

Chapter 5
Coal-fired space heaters

Coal-fired space heaters are crosses between furnaces and stoves. They look like small furnaces, but like stoves, they are normally located in the living area of the house. Their heat output ranges from twenty thousand to one hundred thousand BTU per hour.

The heaters on the market are divided about equally between hand-fired coal/wood units and stoker-fired coal-only units. In the stoker models, the stoker is contained completely within the heater (you shovel coal into the built-in, top-opening, two hundred and fifty pound hopper and it is automatically fed through a partition to the fire). Even so, the largest of these units measures only forty-five inches long, twenty-four inches wide and forty-two inches high.

The stoker-fired units heat almost entirely by convection. The stoker moves the coal into a retort surrounded by the combustion chamber, or heat exchanger. As the fresh air entering the heater picks up heat from the heat exchanger, a large circulating fan forces it out through vents in the sides and top of the heater into the surrounding room. From there it finds its way into the adjoining rooms.

Obviously, this does not make for such even distribution of heat throughout a house as you get with a warm-air furnace, but it is equal to or better than the heat distribution you get from stoves, and it can be improved (as in the case of stoves) by installing small fans or grilled openings high in the walls between rooms. See Chapter 3.

Operation of the stoker heaters, which cost about nine hundred dollars, is almost automatic. A room thermostat turns them on and off as the temperature in the room falls and rises. Draft is controlled by the thermostat, and a limit control and hold-fire timer protect against emergencies (as in central warm-air heating plants). All you have to do is load the hopper with coal and take out ashes and clinkers. In the summer, the circulating fan can be activated manually to keep fresh air moving through the house.

Noise of operation is minimized by design measures taken by the manufacturers. Nevertheless, it is too much to expect that a stoker-fired space heater will be as silent as a stove or hand-fired space heater.

Some of the hand-fired coal/wood space heaters on the market heat primarily by radiation; others by convection. They cost around four hundred dollars. A thermostat mounted on the side controls the primary air supply by opening and closing the damper at the bottom of the unit. Once the firebox is loaded, the heater pretty much runs itself until the next loading. You merely shovel in coal and remove the ashes as you do with a stove.

Several of the radiant heaters, which are generally about the same size as the stoker models, can be enclosed in optional grilled cabinets to improve their appearance and increase protection against burns. Alternatively, you can enclose the heaters in solid steel jackets and duct the heat into an existing heating system. In addition, one manufacturer sells a small domestic water heater that is installed in the combustion chamber and feeds into a storage tank or electric water heater.

The other hand-fired heaters are about half the size of the stoker models. They, too, can be tied into an existing heating system.

Installation

All space heaters are designed for easy installation by the buyer. The only thing you have to do is place the heater and hook it up to an approved all-fuel chimney with a short stovepipe.

Because the combustion chamber in the stoker-fired heaters is surrounded by air, the cabinet is cool enough to touch without getting a burn and you can place the unit within a few inches of combustible surfaces. However, eighteen inch clearance between combustible surfaces and the stovepipe is mandatory, so the vent side of the heater must be placed eighteen inches out from a wall.

The stoker manufacturers do not even call for a floor protector. But we'd prefer one because there is a mild hot spot directly under the retort. A ¼ inch asbestos floor mat covered with metal is adequate. This need not extend beyond the sides of the heater except to protect the floor from spilled coal and ashes.

Clearances required for the hand-fired heaters are about forty inches from all sides to combustible surfaces. The same clearance is necessary above the heater. (Reduced clearances are permissible if the walls and ceiling are protected with noncombustible material.) Under the heaters install a 3/8 inch metal-covered asbestos floor mat. This should extend at least eighteen inches in front of the loading door and eight inches to all other sides.

The smoke pipe for all space heaters is made of twenty-four gauge or heavier steel and installed like that for a stove. See Chapter Three for how to run it through walls and connect it to the chimney.

Operation and maintenance

Follow the directions for stoves or stoker-fired central heating plants.

Made specifically for burning anthracite, this grate has a fan to keep the fire burning vigorously and circulate the heat given off. (Fairclough Coal)

Chapter 6
Burning Coal in fireplaces

Right at the outset we must make it clear that as a heating device, a fireplace leaves an awful lot to be desired. Only a small percentage of the heat it puts out reaches the occupants of the room; most of it goes up the chimney.

This situation can be improved, of course, if you use in the fireplace one of the many boosters that have been put on the market since the oil crisis began. In fact, with one of the grate-type boosters and a small fan attachment, Stanley can raise the temperature of his dining room—a very large room—from below freezing to a comfortable sixty-five to seventy degrees with coal or wood alone. This is impossible with an ordinary fire; he's tried it.

Even so, fireplaces are mighty inefficient heaters. That is as true if you use coal as it is with wood.

Coal is an excellent fireplace fuel. In the days when it was the leading fuel for all kinds of heating, many home owners burned it in fireplaces. But while it has a higher BTU value than wood and burns much longer, it lacks several of wood's virtues: It doesn't have a pleasant aroma when burning; at its worst, it is quite unpleasant. Except for cannel coal, it burns quietly with a low, single-color flame. Because of its byproducts, you shouldn't toast marshmallows over it. And it's dirtier to handle.

Which coal to use

Any coal can be used in a fireplace but cannel coal, despite its premium price, is far and away the best because it gives off flames comparable to those of oak, hickory, apple, even pine. You might call it a "fatty" coal because it burns so exuberantly with a lot of snap, crackle and pop—as if it contained fat. Furthermore, it has a high though variable BTU content. It's very east to ignite. And it comes in large chunks that don't have to be broken up (although you can if you want); therefore, it's easy to handle. Its only real drawback is that, as the coal heats up and burns, it pops off small splinters and chunks and sends them sailing like the ash and cinders from Mt. Saint Helens. If your fireplace is closed with a draw-type screen, the pieces are confined to the fireplace. But if the screen is open, they fly out into the room like the embers from a wood fire. We can't recall that we have ever seen them burning or smoking, so there isn't much danger that they will burn a hole in the floor or start a conflagration. But they do make a fair mess that has to be cleaned up the next day.

Other types of coal are much less satisfactory because they burn quietly and with considerable smoke until they're burning full blast. The softer the coal, the more readily it burns. That means that lignite starts and burns more freely than sub-bituminous; sub-bituminous starts and burns better than bituminous; and bituminous starts and burns better than anthracite. In fact, anthracite almost needs a forced draft of air to keep it going strong.

Heating equipment

You can burn coal in any fireplace—a conventional masonry fireplace; a conventional fireplace made of steel; or a freestanding steel fireplace. But to do so, you should have a coal grate, enclosed on four sides, which holds the coal and fire well above the ashes that fall through to the bottom of the fireplace.

The front-to-back depth of the grate is not very important but the width is. Measure the width of the back wall of the fireplace and get a grate that is just about this wide. If you use a very small grate, the air entering through the fireplace opening will tend to flow around the sides of the grate where it meets no resistance; and the fire will not get enough air to keep it going strong.

Set the grate into the fireplace against the back wall.

This C-shaped heat grate does not require a fan since the fireplace opening is covered, thus preventing the heated air emanating from the top of the tubes from being sucked back up the chimney. The maker supplies a coal basket to fit the grate. (Envirostyle Direct)

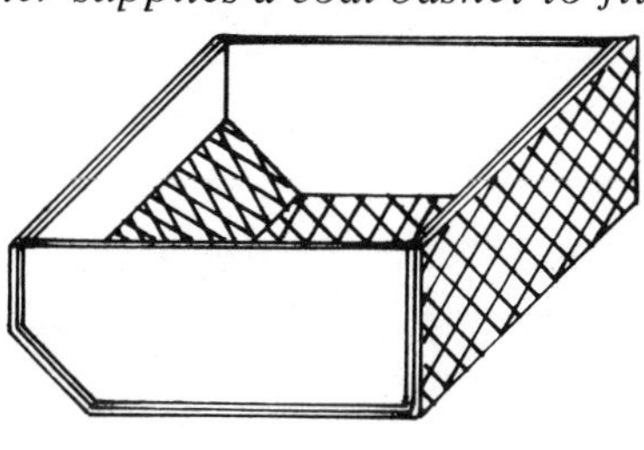

This is a conventional heat grate with a fan that forces air through the tubes and far out into the room. It's designed for a wood fire but can be adapted for coal if you set a coal basket over the tubes. (Stites)

There are several alternatives to the conventional grate. A one-of-a-kind unit made by Fairclough Coal & Oil Co., 800 E. 27th St., Paterson, New Jersey 07513 is a large grate made specifically for anthracite though it can be used with other coals. This actually is more like a stove without a top than a grate. It's a sizable, brick-lined steel box enclosed at the bottom (with a cleanout door in front) and open at the top. A small fan in the side forces air into the box to keep the fire going.

The other alternatives are heat boosters which by various means capture some of the heat that normally goes up the chimney and circulate it into the room. (The most familiar kind of booster is a C-shaped unit, raised off the bottom of the fireplace, with four or five large hollow tubes. As air enters the tubes at the bottom end—either naturally or under forced draft—it is heated and flows out into the room at the upper end.) As we said earlier, we strongly recommend boosters if you want to extract as much heat from a fireplace as possible.

Several boosters that can be used with coal are listed in the back of the book. Although we haven't tried it, we think you can also convert many heat boosters designed for wood to coal use. This looks especially feasible with the C-type booster. All you'd have to do is set an ordinary coal grate into the booster.

Running a coal fire

Be sure the damper is wide open at all times.

To start a coal fire you need plenty of paper and kindling. (We said in Chapter Two that you can light cannel coal with a match, but don't waste time trying.) Add only two or three smallish lumps of coal. The paper and kindling not only quickly heat up the flue so you get a good draft but also ignite the coal.

Once the kindling is burning merrily and the coal lumps are catching, add a few more smallish lumps of coal; then some more. You want to maintain and stimulate the draft, so don't snuff it off by using too much coal or very big lumps at the outset.

As the coal burns along, a thick bed of glowing coals builds up. Add new lumps while the old coals are still active. If the old coals are allowed to cool and become coated with ash, new coals won't catch so easily and the production of volatiles will increase. There's also a chance that some of the new coals will explode, although rarely in a dangerous way. If you let a coal fire burn down too low before adding fresh coal, you run the chance of extinguishing it. To avoid this, put a little paper and kindling under the fresh coal so it will ignite rapidly.

Most of the ash produced falls through the grate, but it doesn't always. So before building a fire, push most of the ash out of the grate (you should keep a thin layer to insulate the grate and prevent possible warping by direct high heat) and remove some or all of it from the bottom of the fireplace so it doesn't impede air movement up through the coal.

Safety

Remember that in a fireplace coal needs a large amount of air to burn well—more than a wood fire does. It also needs this air to carry the dangerous products of combustion up the chimney.

We are sure that what we're about to say will bring some letters saying that it's nonsense. But we stand by our guns.

In a fireplace a coal fire produces more carbon monoxide than a wood fire and therein lies a serious hazard. We can't cite examples to prove that coal fires have killed people but we refer you to the burning of charcoal or charcoal briquettes indoors. Charcoal is much like coal in size and burns much like coal, rather quietly. There have been numerous reported instances of people dying of carbon monoxide poisoning when they broiled meat in a charcoal brazier in a fireplace indoors. The same thing can happen and undoubtedly has if coal is not burned properly in a fireplace.

The way to avoid this is to keep the fire burning well—not smoldering—and to make sure that it is getting plenty of air at all times. If you have a tight house—heavily insulated and weather-stripped—the air that filters in from outdoors is probably inadequate to maintain vigorous combustion; and without vigorous combustion, the carbon monoxide given off by the fire builds up in the room to perhaps hazardous levels. It follows that you must increase the air supply to the fire. The easy way to do this is to crack open a window. The better way is to install a three inch air duct leading from the outdoors under the floor to the fireplace and installing a grilled opening in the floor in front of the hearth or, better still, in the hearth just outside the fireplace opening.

As you get ready to retire at night, keep the fire going strong (without adding fresh fuel) by jarring the ash from the coal. This will hasten extinguishment after you go to bed. Do not close the damper. If you want to keep the heat in the house from escaping up the chimney during the night, seal the fireplace opening either with a tight-fitting built-in glass firescreen or by covering it with a home-made screen. For this you need nothing more than a sheet of

½-inch plywood cut an inch or two wider on all sides than the fireplace opening and a sheet of ¼-inch asbestos-cement board cut to fit snugly within the fireplace opening. Screw the cement board to the back of the plywood; attach a couple of handles to the front of the plywood and, at the bottom, a couple of large angle irons to hold the screen upright when in place. Before going to bed, just fit the screen into the fireplace opening as tightly as possible. This will cut off the air to the fire and it will die quickly.

Grille in hearth at mouth of fireplace opening admits air, brought in from outdoors via a duct, to keep fire—whether wood or coal—supplied with oxygen so it can burn properly.

Even when doors of tight-fitting glass firescreens are closed, sufficient air is admitted through base plate to keep fire burning. When air inlet is closed at night, fire dies out quickly and screen prevents warm room air from escaping up the chimney.

Chapter 7
Living with coal heat

Coal availability

We've indicated here and there in preceding chapters that the availability of coal through coal dealers in your area is not assured. But this situation is changing. For example, to buy coal by the ton, home and store owners in our part of the world must order it from at least thirty miles away. But there are various nearby stores and entrepeneurs who sell it in small quantities; and just recently one of these reported that his sales rose so dramatically last winter that he is seriously considering opening a coal yard to sell by the truckload.

So we're sure that coal will be generally available in reasonably heavily populated areas in more and more parts of the country fairly soon. But this still doesn't answer all questions:

Is it the kind of coal you need? Is the price fair? Does the dealer deliver it right into your bin?

You can find the answers only by making the rounds of the coal dealers.

If you want hard coal and the dealers sell only soft coal (or vice versa), that is no reason to give up on coal heating. You just have to buy a stove, furnace, etc. that burns soft coal.

Whether the price is right is a matter you alone can decide after comparing the cost of coal heating with that of your present heating. One thing to bear in mind is that the price a dealer charges depends on the amount of work he has to do to get the coal to you. If you drive a truck to the coal yard and load it yourself, you'll pay a minimum price. If you want the dealer to deliver the coal to your home, you pay a considerably higher price. How much higher depends on whether the coal can be dumped directly from the truck into the bin or whether it has to be carted in in bags or wheelbarrows.

Coal delivery practices vary at the moment because there are so few dealers that competition has not forced them to cater to the consumer. Some won't deliver coal right into the bin unless they can chute it in from the truck. In this case, you either go without or have the dealer dump it in your driveway and you carry it in yourself. All other dealers charge extra for having their men wheel or lug the coal from the truck to the bin. While we haven't run into any dealers who do not deliver coal at all—leave it to you to pick it up at the yard—there probably are such. On the other hand, there seem to be a good many dealers who will not make deliveries in less than two, three, four or five-ton lots.

As we said, these are things you have to find out. The only thing you can be certain of is that, unlike oil and bottled-gas dealers, coal dealers do not make automatic deliveries. When you see that your bin is running dry, you have to call the dealer and say, "Send me more."

Coal storage

Coal can be stored outdoors as long as you keep the pile dry with huge plastic sheets or tarpaulins. But who wants a coal pile in the backyard?

Safest storage is indoors. This means building a bin to keep the coal in one place. There is nothing much to do after you've settled on the dimensions of the bin (a 2 x 4 x 5-foot space for each ton).

To save expense and work, put the bin in a corner of the basement under one of the windows. Use 2 x 4's and ½-inch plywood or one inch boards (preferably tongued and grooved). First, fasten 2 x 4-inch soleplates to the floor. Since the floor is probably concrete, drill holes spaced no more than two feet apart in it and tap in lead anchors. Then drive screws through the plates into the anchors. Applying thick squiggles of construction cement to the bottoms of the 2 x 4's before screwing them down adds to the strength of the walls.

Directly above the soleplates install 2 x 4-inch top-plates. Actually, if you lay the bin out properly, you'll need only one of these. It runs at right angles to the ceiling joists and is spiked to the bottoms of the joists. The top-plate on the other side of the bin is one of the ceiling joists.

Now cut the necessary number of 2 x 4's to fit snugly between the soleplates and top-plates. These are the studs. You need two studs nailed together on both sides of the coal-bin door; a single stud at the foundation-wall end of each coal-bin wall. In between, the studs should be spaced twelve to no more than sixteen inches apart. Toenail all the studs securely to the plates with 3½-inch common nails. Use four nails in the bottom of each stud and three or four nails in the top.

The plywood sheets or boards for the walls are nailed to the inside edges of the studs and soleplates. This is better than nailing them outside because the pressure of the coal cannot force them away from the studs and you eliminate pockets between studs in which some coal will be trapped.

The plywood or board sides need not extend all the way up to the ceiling since you can't load in coal that high; but they should go up far enough so coal cannot fall out as it's loaded in.

The opening into the bin should be at least twenty-four inches wide. Locate it where it's easy to shovel out the coal and toss it with minimum effort into the heating plant. The door is made of 1 x 8 or 1 x 10-inch square-edged boards that slide up and down in channels attached to the inside walls of the bin. Each of the two channels required can be made of 1 x 2's nailed together to form an "L," and then nailed through the bin walls into the double studs to form a "U." The boards for the door are then cut about two inches wider than the door opening and dropped down into the channels. The bottom board should be about eight inches above the floor so you can slide in your coal shovel and bring out a load of coal. Whenever you have to get into the bin to open the basement window or pull the coal forward in the bin, just slide the boards up and out of the channels.

A bin for a bin-feed stoker is constructed in the same way except that the door is closed all the way down to the floor; and the tube enclosing the feed screw is inserted through a wall of the bin or the bottom board in the door.

Since moisture in coal may cause spontaneous combustion, you should always ask for delivery on a dry day. (Piling wet coal on top of dry coal is especially dangerous.) Do not let the dealer spray the the coal after delivery to reduce dust: the dust will settle quickly enough and you don't want to run the risk of fire. If your basement is very damp or humid, air it out well from time to time. If you notice that the coal pile is getting hot, call the fire department at once.

There are several other precautions to take in managing coal storage

Keep the bin free of leaves, straw, rags, wood scraps, etc.

Try to hold the temperature in and around the bin to under seventy-five degrees. Warm-air ducts, hot-water pipes and steam pipes should not run through the bin.

Don't let children play in the bin. You don't want them to get filthy. Most of all, you don't want them to become buried in the coal—not an unknown catastrophe.

Ash disposal

The soft ashes from a coal fire are worth absolutely nothing. The hard cinders and clinkers, when broken up, make good ballast (in place of gravel or crushed stone) for walks, driveways, terraces, etc. But there's a limit to how many of them you can use.

Another possible way to use small hard cinders is to dig them into clay soil to make it more penetrable by roots and water. But you must go easy here because the cinders are acid and will upset the soil pH.

You can also use small, gritty ash as a substitute for sand on iced-over walks and driveways, but they make a mess and are murderous to floors when tracked into the house.

Really, then, there isn't much use for coal ash, and the best thing is to send or take it to the town dump.

Pending pick-up by the trash collector, the ashes should be kept in sound metal ash cans with tight-fitting lids to snuff out any hot coals you dredge from the heating appliance. Keep the cans away from combustible materials. After the ashes are cool, if you want to try to sift out any good coals that have fallen through the grate, go ahead. But this should be hardly worth the mess and effort. We say "should be" because if you salvage too much coal, it's a sign that you're not running the furnace, stove, etc., properly.

Chimney fires

If you've never had or have never seen a chimney fire, take it from those who have: the noise alone will scare the bravest man out of his wits. The fire may not ignite the house but yet again it may. And it may very well damage the chimney, thus necessitating expensive reconstruction.

Obviously, the best ways to avoid chimney fires are (1) to keep the chimney clean; (2) to forbid burning of trash in the fireplace or heater; and (3) to resist over-firing, which is done by piling on too much coal and/or opening the drafts too wide.

But suppose, despite all precautions, you do have a chimney fire. How do you bring it under control?

1. Alert the household and have someone summon the fire department immediately. But if you're by yourself, take Step 2 before calling the fire department.

2. Close off the draft. This is simple enough to do if the heater is airtight; not so easy if it isn't. But do what you can anyway by closing all air inlets.

If the heater is vented into a flue that serves another heating device, don't forget to shut off the drafts in the latter, too. However, if the second heating device is an oil or gas-fired furnace or water heater, you must turn it off before closing its drafts; otherwise, it may automatically come on while the draft is blocked.

Shutting off the air to a fireplace is more difficult. If it's equipped with a glass firescreen, close the doors and air inlet. Even so, these screens rarely fit so tightly in the fireplace opening that air cannot enter around the edges. Fireplaces without solid screens can be blocked off by holding a large sheet of metal or asbestos-cement board over the opening or draping a wet blanket or mattress pad over the opening. Just be careful that the suction in the chimney doesn't draw a blanket into the fire.

3. Try to stifle the fire in the fireplace, etc. One way to do this—though it may not have as much effect on the fire in the chimney as you'd like—is to scatter a large box of baking soda on the flames. A better method is to use a chimney-fire extinguishing flare according to the maker's directions. These are sold by the Signal Flare Division of Standard Railway Fuse Corp., Box 178, Boonton, New Jersey, 07005. A dry-chemical fire extinguisher can also be used, but you actually need two of them for real effect. Use one to extinguish the fire on the grate; the second to direct up the chimney to snuff out the chimney fire.

Water should be used very judiciously lest the sudden cooling of the heater and chimney causes them to crack. Sprinkling it on the fire in the heater in half cupful quantities is enough to discourage the flames.

4. Until the firemen arrive, put out the fire and check the house, keep a watch for the spread of the fire from the chimney to the neighboring structure. The walls around the chimney may be ignited by the intense heat or by flames licking through cracks in the flue lining. The roof may be ignited by sparks issuing from the chimney.

Cleaning a chimney

Although coal doesn't create creosote as wood does, it does create soot and flyash which might catch fire and which will surely affect draft and gradually stifle combustion in the stove, furnace, etc. For these reasons, you should inspect the chimney by looking up or down it with a flashlight at least once a year; and you should clean it whenever the buildup of soot seems excessive.

Since chemical cleaners that are tossed into a fire are not very effective and may cause corrosion, the best way to clean a chimney is by getting up on the roof and brushing it thoroughly from top to bottom. The ideal tool is a steel-bristled or fiber-bristled brush made for the purpose. These come in various sizes, round or rectangular, to fit most flues. Get one of the same dimensions as the flue.

Before starting work, make sure the cleanout door (if there is one) at the base of the chimney is closed. In a fireplace, open the flue damper and tape a sheet of plastic, cardboard or plywood over the fireplace opening.

Then climb up on the roof and make sure you're safe. Cleaning brushes come with a rope loop on one end and a pipe thread on the other. To lower the brush into the chimney, either screw on threaded metal rods or tie a rope to one end and a ten pound chunk of iron to the other end. Run the brush up and down inside the flue four or five times to scour off the soot.

When you're satisfied that the chimney is clean, wait a few minutes while the dust settles. Then remove the soot from the fireplace or cleanout or, if you don't have a cleanout, from the chimney connector (which must be disassembled).

Other devices for chimney cleaning include small Christmas trees, burlap sacks stuffed with leaves and weighted with bricks, or tire chains. Of these, the chains are most effective and, except for brushes, they're the only thing you can use in chimneys in which the flue bends. Just don't slap the chains so hard against the flue lining that you crack it.

However and whenever you clean your chimney, take the occasion to clean the chimney connector also.

Noise

No doubt about it, shoveling coal and shaking down and removing ashes from a furnace and boiler make noise that you can hear in the rooms directly overhead. A stoker also makes noise when it's feeding coal into the combustion chamber. We personally do not think this is much to fret about. The loudest noises are made by hand-firing operations but are generated only in the morning and evening. Stoker noise is muffled. If it isn't, you or the dealer had better have a look at the machine.

Modern oil-fired furnaces and boilers often make far more noise than coal-fired units. They run with a rumbling roar you can hear all over a house.

However, if you are bothered by coal-heating noises, you can stifle them pretty well by nailing two layers of 5/8-inch Type X gypsum board to the bottom of the joists in the basement.

Dirt

If your coal heater (whichever type it may be) is tight and in good condition and you run it properly, it won't stir up any more dust in your house than any other heater. The smoke pipe must also be tight and sound.

But of course you will create dirt when stoking and cleaning a coal heater. Most of this is concentrated in the immediate vicinity of the heater. To keep it down, just don't load, shake or remove ashes too violently. Then get out a broom or vacuum cleaner.

In the basement, you won't be so careful; and in time you'll have a film of coal dust all over the place. Vacuuming is the best way to get it up if you can persuade the lady of the house to let her vacuum be used. If you use a broom, use along with it an oily sawdust-like cleaning compound such as janitors use in schools. When sprinkled on the floor and swept up, this helps to coagulate and pick up the dust and keep it from drifting. You can reuse the same dirty compound time and again until it clearly can absorb no more dirt.

Chapter 8
Who makes coal-heating equipment

Stoves

A-1 Electric & Plumbing
Box 66117
Portland, Oregon 97266
"Acme"–airtight

All Nighter Stove Works, Inc.
80 Commerce Street
Glastonbury, Connecticut 06033
"Chubby Moe"–airtight

Atlanta Stove Works, Inc.
Box 5254
Atlanta, Georgia 30307
Radiant, circulating, cookstove

Bart Manufacturing Company
5019 Hovis Road
Charlotte, North Carolina 28208
"Black Bart"–circulating, fireplace insert

Birmingham Stove & Range Company
Box 2647
Birmingham, Alabama 35202
Radiant, circulating, cookstove

Bow & Arrow Stove Company
14 Arrow Street
Cambridge, Massachusetts 02138
"Preporod", "Chappee", "Le Petit Godin", "Supra"– airtight circulating, cookstove

Ceramic Radiant Heat
Pleasant Drive
Lochmere, New Hampshire 03252
"Weso"–ceramic airtight

Classic Stove Works
233 Main Street
New Britain, Connecticut 06051
"Surdiac"–airtight

Comforter Stove Works
Box 175
Lochmere, New Hampshire 03252
Airtight

Crane Stove Works, Inc.
Box 440
Braintree, Massachusetts 02184
Airtight

Enterprise Foundry Company
Sackville, New Brunswick EOA 3CO
Cookstove

Finest Stove Imports, Inc.
Box 1733
Silver Spring, Maryland 20902
"Koppe"–ceramic airtight

Fisher Stoves International, Inc.
Box 10605
Eugene, Oregon 97440
"Coal Bear"–airtight

Franco–Belge
Franco-Belge of America
70 Pine Street
New York, New York 10005
Airtight

Glacier Bay
6743 Kinne Street
East Syracuse, New York 03057
Airtight, fireplace insert

Good Time Stove Company
Route 112
Goshen, Massachusetts 01032
"Lunenberg"–radiant

Heathdelle Associates
RFD 4, Box 153
Laconia, New Hampshire 03246
"Nashua"–airtight

KNT, Inc.
Box 25
Hayesville, Ohio 44838
"Impression"–circulating

LaFleur Distributing
4629 Clyde Park S.W.
Grand Rapids, Michigan 49509
"Esse Dragon Mark II", "Dolphin"–airtight

Locke Stove Company
114 W. 11th
Kansas City, Missouri 64105
"Warm Morning"–circulating

Malleable Iron Range Company
715 N. Spring Street
Beaver Dam, Wisconsin 53916
"Monarch"–radiant, circulating, cookstove

Martin Industries
King Products Division
Box 128
Florence, Alabama 36530
"King"–radiant, circulating

The Merry Music Box
20 McKown Street
Boothbay Harbor, Maine 04538
"Styria"–airtight, cookstove

Mohawk Industries
121 Howland Avenue
Adams, Massachusetts 01220
"Temp Coal"–airtight

National Stove Works, Inc.
78OH Howe Caverns Road
Cobleskill, New York 12043
"Thermo-Control"–airtight

Old Country Appliances
Box 330
Vacaville, California 95688
"Tirolia"–circulating, cookstove

Old Erie Stoves, Ltd.
7000 Fly Road
East Syracuse, New York 13057
"Glacier Bay"–airtight, fireplace insert

Scandinavian Stoves, Inc.
Box 72
Alstead, New Hampshire 03602
"Lange"–airtight

Shenandoah Manufacturing Company
Box 839
Harrisonburg, Virginia 22801
Airtight, circulating

Suburban Manufacturing
Box 399
Dayton, Tennessee 37321
Circulating, fireplace insert

Superior Stove Company
Box 218
Stoughton, Massachusetts 02072
"Flame View", "Hot Shot"–airtight, fireplace insert

United States Stove Company
South Pittsburg, Tennessee 37380
Radiant, circulating

Vermont Castings, Inc.
Prince Street
Randolph, Vermont 05060
"Defiant", "Vigilant"–airtight

Washington Stove Works
Box 687
Everett, Washington 98201
Radiant, cookstove

Hand-fired furnaces and boilers

Type of coal is given only when indicated in manufacture literature.

Aquagen, Inc.
365 E. Cedar Street
Newington, Connecticut 06111
Coal/wood boiler—anthracite

Beck Engineering Combustion Company
2100 Cole Street
St. Louis, Missouri 63106
Coal-only furnaces—forced air, bituminous

Dornback Furnace & Foundry Company
33220 Lakeland Boulevard
Eastlake, Ohio 44094
Coal/wood furnaces—forced air and gravity, anthrac

Duo-Matic Olsen, Inc.
2510 Bond Street
Park Forest South, Illinois 60466
Coal/wood furnace—forced air
Multi-fuel furnaces—forced air

Electric Furnace-Man Division
General Machine Corporation
Emmaus, Pennsylvania 18049
Coal/wood furnace—forced air

Heathdelle Associates
(Distributor for Trianco)
RFD 4, Box 153
Laconia, New Hampshire 03246
Coal-only boiler—anthracite

Heat Shed, Inc.
(Distributor for HS-Tarm)
Rock Ridge Road
Revere, Pennsylvania 18953
Coal/wood boilers
Multi-fuel boilers

Malleable Iron Range Company
Beaver Dam, Wisconsin 53916
Coal/wood furnace—forced air

Newmac Manufacturing, Inc.
Box 545
Woodstock, Ontario N4S 7Y5
Multi-fuel furnace—forced air
Multi-fuel boiler

New Yorker Steel Boiler Company
Colmar, Pennsylvania 18915
Coal/wood boiler—anthracite

Oneida Heater Company
Box 148
Oneida, New York 13421
Coal-only furnaces—forced air and gravity
Coal/oil furnaces—forced air
Coal/wood boiler
Multi-fuel boilers

Powrmatic, Inc.
2906 Baltimore Boulevard
Finksburg, Maryland 21048
Coal/wood furnaces—forced air
Multi-fuel furnaces—forced air

Riteway Manufacturing Company
Box 153
Harrisonburg, Virginia 22801
Coal/wood furnaces—forced air
Multi-fuel furnaces—forced air
Coal/wood boilers
Multi-fuel boilers

S/A Energy Division
4642 Crossroads Park Drive
Liverpool, New York 13088
Coal/wood boiler
Multi-fuel boiler

Slant/Fin Corporation
100 Forest Drive
Greenvale, New York 11548
Coal/wood boilers—anthracite or bituminous

Suburban Manufacturing
Box 399
Dayton, Tennessee 37321
Coal/wood furnace—forced air

Thermo Products, Inc.
North Judson, Indiana 46366
Coal/wood furnaces

Van Wert Manufacturing Company
739 River Street
Peckville, Pennsylvania 18452
Coal/wood boiler
Multi-fuel boiler

Stoker-fired furnaces and boilers

Type of coal is given only when indicated in manufacturer's literature.

Axeman-Anderson Company
233 West Street
Williamsport, Pennsylvania 17701
Boilers–anthracite

Beck Engineering Combustion Company
2100 Cole Street
St. Louis, Missouri 63106
Furnaces–bituminous

Electric Furnace-Man Division
General Machine Corporation
Emmaus, Pennsylvania 18049
Boiler–anthracite
Coal/oil boiler–anthracite

Gentleman Janitor Manufacturing Company
Scranton-Carbondale Hwy.
Route 6
Scranton, Pennsylvania 18508
Boilers–anthracite

Keystone Manufacturing Company
Schuylkill Haven, Pennsylvania 17972
Furnaces—anthracite
Boilers—anthracite

Prill Manufacturing Company
Drawer S
Sheridan, Wyoming 82801
Furnaces—sub-bituminous and lignite
Boilers—sub-bituminous and lignite

Slant/Fin Corporation
100 Forest Drive
Greenvale, New York 11548
Boilers—anthracite (stokers supplied by others)

H. B. Smith Company
Westfield, Massachusetts 01085
Boilers for large homes and commercial establishments (stokers supplied by others)

Southern Schuylkill Manufacturing Corporation
Orwigsburg, Pennsylvania 17961
Boilers
Furnaces

Stokermatic Corporation
1610 Industrial Road
Salt Lake City, Utah 84104
Furnace

Van Wert Manufacturing Company
739 River Street
Pecksville, Pennsylvania 18452
Boilers—anthracite

Will-Burt Company
Orrville, Ohio 44667
Furnace

Stokers for installation in hand-fired furnaces

Losch Energy Systems, Inc.
Box 1171
Pottsville, Pennsylvania 17901

Prill Manufacturing Company
Drawer S
Sheridan, Wyoming 82801

Will-Burt Company
Orrville, Ohio 44667
Stoker-fired
Hand-fired

Space heaters

Dornback Furnace & Foundry Company
33220 Lakeland Boulevard
Eastlake, Ohio 44094
Hand-fired

Prill Manufacturing Company
Drawer S.
Sheridan, Wyoming 82801
Stoker-fired

Riteway Manufacturing Company
Box 6
Harrisonburg, Virginia 22801
Hand-fired

Stokermatic Corporation
1610 Industrial Road
Salt Lake City, Utah 84104
Stoker-fired

Will-Burt Company
Orrville, Ohio 44667
Stoker-fired
Hand-fired

Fireplace heat boosters

Aquappliances, Inc.
135 Sunshine Lane
San Marcos, California 92069

Belmont Industries, Inc.
155 John Street
Reading, Massachusetts 01867

Envirostyle Direct
Box 43021
St. Paul, Minnesota 55164

Lance International
199 Forest Street
Manchester, Connecticut 06040

Stites Manufacturing Company
615 Hunter Lane
Santa Rosa, California 95404

Thermalite Corporation
Box 658
Brentwood, Tennessee 37027

Woodmack Products, Inc.
850 Aldo Avenue
Santa Clara, California 95050

Chimney heat reclaimers

Chimney Heat Reclaimer Corporation
53 Railroad Avenue
Southington, Connecticut 06489

Dolin Metal Products, Inc.
475 President Street
Brooklyn, New York 11215

Electric Furnace-Man
General Machine Corporation
Emmaus, Pennsylvania 18409

Fuel Saver
42 W. Blaine Street
McAdoo, Pennsylvania 18237

Lance International
199 Forest Street
Manchester, Connecticut 06040

Magic Heat Corporation
Burton, Michigan 48529

J. H. Stoneberg Manufacturing Company
Brockport, Pennsylvania 15823

Torrid Manufacturing Company
1248 Poplar Place, South
Seattle, Washington 98144

Prefabricated insulated chimneys

Dura-Vent Corporation
2525 El Camino Real
Redwood City, California 94064

Enterprise Sales
RFD 2
Apple Creek, Ohio 44606

General Products Company
Box 887
Fredericksburg, Virginia 22401

Hart & Cooley Manufacturing Company
Box 903A
Holland, Michigan 49423

Oliver-MacLeod, Ltd.
5 Edward Street
Gravenhorst, Ontario POC 1GO

Wallace-Murray Corporation
1820 East Fargo
Nampa, Idaho 83651

Ceiling fans

Acorn Machinery Company
182 Lafayette Street
New York, New York 10013

Airmaster Fan Company
150 West North Street
Jackson, Michigan 49204

Airtrol Corporation
204 Harvick Street
Rockton, Illinois 61072

American Way Farm Products
Route 403 South
Johnstown, Pennsylvania 15907

Casa Blanca Fan Company
182 South Raymond Avenue
Pasadena, California 91105

Chelsea Fans
Tuttle & Bailey
215 Warren Street
New Britain, Connecticut 06052

Cincinnati Fan & Ventilator Company
5317 Creek Road
Cincinnati, Ohio 45242

Delaware Electric Imports
111 South Delaware Avenue
Yardley, Pennsylvania 19067

Electric Trading Company
1 Mercer Street
New York, New York 10013

Emerson Electric Company
8100 West Florissant Avenue
St. Louis, Missouri 63136

Energy Conservation Systems
189 West Newberry Road
Bloomfield, Connecticut 06002

Frigid, Inc.
1250 Rockaway Avenue
Brooklyn, New York 11236

Great Things
Box 450271
Miami, Florida 33145

Gulf Coast Fans
1771 Sunshine Drive
Clearwater, Florida 33515

Hamilton Fan & Blower Company
5347 Creek Road
Cincinnati, Ohio 45242

Hunter Division
Robbins & Myers
2500 Frisco Avenue
Memphis, Tennessee 38114

Krenz & Company
Carnegia Drive
Germantown, Wisconsin 53022

Olde Tyme Ceiling Fan Company
22858 Ventura Boulevard
Woodland Hills, California 91364

Patton Electric Company
11401 Bluffton Road
Fort Wayne, Indiana 46809

Penn Ventilator, Inc.
11th & Allegheny
Philadelphia, Pennsylvania 19133

Phil Rich Fan Manufacturing Company
1001 West Loop North
Houston, Texas 77055

Reynolds Electric Company
1810 Madison Avenue
Maywood, Illinois 60153

Stanley Industrial Corporation
1218 East 9th Street
Jacksonville, Florida 32206

Uno Ventilator Company
1236 Eastern Avenue
Malden, Massachusetts 02148

GODIN

Index